Hunter, Killer

The $200 War

How Cheap Drones Broke Modern Warfare

C. Scott Lannon

While the author has made every effort to ensure the accuracy and completeness of the information in this book, the author and publisher assume no responsibility for errors, inaccuracies, omissions, or any inconsistency herein. The views expressed are the author's own and do not represent the official positions of any government, military service, or defense organization. No classified information was used in the preparation of this work. All sources are publicly available and cited in the selected bibliography.

Published by C. Scott Lannon

ISBN: 978-1-972731-11-6 (Paperback) ISBN: 978-1-972731-10-9 (Digital)

Introduction: Do the Math

On November 9, 2023, a volunteer drone operator sat in a basement near the Kupiansk front in eastern Ukraine, FPV goggles over his eyes, thumbs resting on a controller that cost less than a video game console. His call sign was Grim. He was twenty-six years old, a former IT technician from Dnipro who had joined a territorial defense unit eight months earlier and discovered he had a talent for flying things into other things. On the small screen inside his goggles, rendered in the green-white glow of a thermal camera, he could see the outline of a Russian 9K330 Tor surface-to-air missile system — a mobile fortress of phased-array radars and vertically launched missiles designed to shield armored columns from aerial attack. The Tor-M2, in the configuration visible on Grim's screen, cost the Russian Ministry of Defense approximately $24 million.

The drone streaking toward it had been assembled three days earlier in a workshop sixty kilometers behind the front line. Its motor was a brushless unit sourced from a DJI supplier. Its frame was carbon fiber, cut on a CNC router donated by a tech company in Lviv. Its tail fin was 3D-printed. Its warhead was a repurposed RPG round, fitted with a 3D-printed adapter sleeve that a volunteer engineer had designed on his laptop and uploaded to a shared Telegram channel. The entire drone, warhead included, cost less than dinner for two in Kyiv. Less than $200.

Grim pushed the stick forward. The thermal signature of the Tor filled his screen — the boxy launcher growing larger, the heat plume from its generator visible as a white smear against the cold ground. Impact. The feed went black. In the basement, no one spoke.

Then someone did the math.

One Tor system could buy 120,000 of the drones that had just destroyed it. The entire Russian air defense battery — four Tor launchers and their associated radar and command vehicles — represented a capital investment roughly equal to Ukraine's total FPV drone production for a month. A volunteer with three days of flight training, using a weapon funded by a civilian crowdfunding campaign, had just erased $24 million in military hardware. Somewhere in the economics of warfare, something had cracked in half.

The footage from Grim's camera, pulled from the drone's onboard recorder before the strike and from a second observation drone hovering overhead, appeared on Ukrainian military Telegram channels within hours. Within days, it had been analyzed by defense commentators across three continents. The technical details mattered — the angle of approach, the warhead placement, the thermal signature exploitation — but they were not what made the video significant. What made it significant was the price tag. Defense analysts at the Royal United Services Institute (RUSI) in London, at RAND in Santa Monica, and at the Center for Strategic and International Studies in Washington, D.C. all converged on the same observation. The cost of the weapon that destroyed the Tor was less than the cost of the fuel the Tor had burned while idling.

The Kupiansk strike was not unique. By late 2023, Ukrainian FPV operators were conducting hundreds of similar attacks every week — cheap, expendable, one-way drones flown by volunteers with weeks of training, destroying equipment that had taken years and millions of dollars to build. The footage accumulated on social media and OSINT channels until it formed a dataset that no military planner could ignore. The individual strikes were tactically significant. The aggregate was strategically transformative. The economics of warfare had not merely shifted. They had inverted.

The Number That Rewrites Everything

This book is about a price. Not a technology. Not a strategy. Not a doctrine. A price: $200.

That number — the approximate cost of a first-person-view drone assembled from commercial components — is the most consequential fact in modern warfare. It is more important than any weapons program, any defense budget, any alliance structure, because it has inverted the fundamental economic equation that has governed armed conflict for centuries. For as long as wars have been fought, the side with the more expensive weapons has held the advantage. Superior technology, superior training, superior platforms — these cost more, and they were worth more. The $200 drone has reversed that logic.

In *Hunter, Killer*, I mapped the full revolution — the technology, the ethics, the policy, the battlefield. This book follows the money. Every chapter delivers one economic argument, one story that proves it, and one number the reader will not forget. Because the drone revolution is, at its core, an economic revolution. And the economics are devastating.

The numbers span every domain of warfare. On land, a $200 FPV drone destroys a $24 million Tor air defense system — a cost-exchange ratio of 120,000 to 1. At sea, Ukrainian Magura V5 sea drones costing roughly $250,000 each have struck Russian warships valued at $65 million or more — a ratio of 260 to 1. In the Red Sea, Houthi forces fire $35,000 Shahed-pattern drones and anti-ship missiles at US Navy destroyers, which respond with SM-6 interceptors costing approximately $4.9 million apiece — a ratio of more than 100 to 1 in the Houthis' favor. And in the most absurd expression of the inversion, a $200 hobbyist-grade surveillance drone loitering over an air defense battery can force the launch of a $4.2 million Patriot missile — a ratio of 21,000 to 1.

These are not anomalies. They are the new arithmetic of conflict. And they all share a common denominator. The weapons doing the damage cost orders of magnitude less than the weapons they destroy, and the gap is widening.

The Cost-Exchange Ratio

Military strategists have a term for this arithmetic: the **cost-exchange ratio**. It measures the relationship between what an attacker spends and what a defender loses. When a $10 million missile destroys a $100 million target, the ratio favors the attacker at 10 to 1. When a $200 drone destroys a $24 million air defense system, the ratio favors the attacker at 120,000 to 1. The cost-exchange ratio is not a new concept. Cold War analysts at RAND and the Office of Systems Analysis used it to evaluate everything from anti-ballistic missile systems to nuclear warhead counts. What is new is the magnitude of the inversion.

In every previous military revolution — the longbow against the armored knight, the torpedo against the battleship, the improvised explosive device against the mine-resistant vehicle — the cost-exchange ratio favored the cheap side by single-digit or low double-digit multiples. A longbow archer cost roughly one-fiftieth of a mounted knight. A torpedo cost roughly one-hundredth of a battleship. Those ratios were enough to reshape entire eras of warfare.

The ratios emerging from Ukraine, the Red Sea, and the broader drone revolution are not in the single digits or the double digits. They are in the tens of thousands. A 120,000-to-1 cost-exchange ratio does not merely favor the attacker. It renders the defender's investment mathematically absurd. No production rate, no maintenance budget, no procurement cycle can sustain a force structure where a single hit from a $200 weapon erases $24 million in capital. The math does not bend. It breaks.

And the math is getting worse. As FPV drone production scales — Ukraine manufactured approximately 2 million drones in 2024 and set a production target of more than 4 million for 2025, with ambitions of 7 to 10 million by 2026 — unit costs are falling further. As AI-assisted targeting improves hit rates, the expected cost per kill drops. As electronic warfare forces drones toward greater autonomy, the human operator's salary — the last significant cost above the hardware — begins to disappear from the equation. The cost-exchange ratio is not static. It is a curve, and it bends in one direction: toward the cheap side.

What This Book Argues

The argument of *The $200 War* is simple, and it is uncomfortable: the economics of warfare have inverted, and the inversion is permanent. Cheap, expendable, mass-produced autonomous weapons now hold a structural advantage over expensive, exquisite, low-quantity legacy platforms. This is not a prediction about the future. It is a description of the present. The evidence is on the battlefield in Ukraine, in the shipping lanes of the Red Sea, in the defense budgets of 40 nations that have acquired armed drones in the past decade, and in the $49 billion that venture capitalists poured into defense technology in 2025 alone.

The book proceeds in four parts. Part I — The Price Tag — examines the cost inversion as it plays out right now, from Ukraine's attrition math to the Houthis' billion-dollar blockade to the Shenzhen supply chain that makes it all possible. Part II — The Pattern — steps back to show that the current inversion is not unprecedented, because every major military revolution has been a cost revolution, and the expensive side has always lost. Part III — The Money — follows the defense dollars, from the Pentagon's $892 billion budget to the counter-drone casino to the US-China drone race. Part IV — The World After — examines what happens when the weapons are so cheap that anyone can afford them, from state actors to garage workshops to autonomous systems that no longer need a human operator.

Every chapter opens with a person in a place at a moment. Every claim is sourced. Every number is specific. The selected bibliography carries the evidence for readers who want to verify, organized by chapter for readers who want to go deeper.

The argument is not that expensive platforms have no value. Aircraft carriers project power. Stealth fighters penetrate defended airspace. Nuclear submarines enforce deterrence. The argument is that the economic foundation on which those platforms rest — the assumption that their cost is justified by their survivability and their capability — is eroding, and the erosion is accelerating. When the cost of threatening a $13 billion carrier drops to the price of a used car, the math that justified building the carrier in the first place demands reexamination. Not abandonment. Reexamination. The distinction matters, and this book will maintain it.

The Stakes

The stakes of the $200 war are not abstract. The United States spends $892.6 billion a year on defense — and the vast majority of that spending is structured around platforms that a $200 drone can threaten. Aircraft carriers, main battle tanks, fourth- and fifth-generation fighter aircraft, guided-missile destroyers — the capital assets of American military power were designed for a world where the cost-exchange ratio favored the expensive side. That world no longer exists.

The question is not whether the Pentagon will adapt. The question is whether it will adapt before a crisis forces the adaptation under fire — in the Taiwan Strait, in the Persian Gulf, in a theater where the cost of learning the lesson is measured in ships and lives rather than budget hearings and white papers.

The same question applies beyond the United States. NATO allies whose force structures mirror America's — heavy on expensive platforms, light on attritable mass — face the same cost-curve exposure. China, which builds both the $3 motors and the military drones that use them, sits on the opposite side of the equation. Russia, which entered Ukraine expecting a three-day war and instead discovered that $200 drones could stall a $20 billion armored advance, has already begun restructuring its military production around the lesson. Every military on Earth is being forced to confront the same arithmetic, and the arithmetic does not care about doctrine, tradition, or institutional preference. It cares about price.

A $200 drone. A $24 million missile system. Do the math. That math is rewriting the rules of war.

But one strike does not make a revolution. What does the math look like when you scale it across an entire war — when both sides are sending thousands of $200 drones at each other every day, and the cost of attrition starts to accumulate? The answer is playing out right now, on the most consequential battlefield in a generation.

Chapter 1: The Attrition Math

In early 2025, Oleksii Babenko walked a production line that had not existed three years earlier. The founder and CEO of Vyriy — now one of Ukraine's largest FPV drone manufacturers — moved between rows of assembly stations in a facility somewhere in western Ukraine, the exact location withheld for operational security. Before him, technicians in anti-static gloves soldered electronic speed controllers to flight boards, fitted brushless motors into carbon fiber frames, and loaded firmware onto processors the size of a thumbnail. Each station produced a complete FPV drone in under forty minutes. Each drone cost a few hundred dollars. Each was designed, from the first circuit trace to the last zip tie, to be used exactly once.

Before Russia's full-scale invasion in February 2022, Ukraine had seven drone manufacturers. By early 2025, it had more than 500. Vyriy alone had invested $4 million in defense-technology acquisitions, absorbing smaller workshops and component suppliers to build a vertically integrated production chain. The drones rolling off Babenko's line were made entirely from domestically sourced components — a deliberate choice to reduce dependence on the Chinese supply chain that fed most of the world's FPV production. They operated on low-frequency channels below 1 GHz, harder for Russian electronic warfare systems to jam. And they were being fitted with AI-assisted targeting through a system called TFL-1, which allowed the drone to track a moving target through its final approach even if the operator's video link was degraded.

Babenko's production numbers reflected a national mobilization without precedent in the history of drone warfare. Ukraine manufactured approximately 2 million drones in 2024 and set a production target of more than 4 million for 2025 — figures that would have been dismissed as fantasy in 2022. The government allocated $2.5 billion to drone procurement and production for the year. The Army of Drones program, launched by Minister of Digital Transformation Mykhailo Fedorov, channeled civilian crowdfunding into military drone units. The Brave1 defense technology incubator fast-tracked startups from prototype to production in weeks. The Diia City economic zone offered tax incentives to defense-tech companies willing to scale. Ukraine was not merely building drones. It was building an entire industrial ecosystem around the premise that cheap, expendable weapons could hold the line against a conventional military ten times its size.

Russia was doing the same thing. The Russian Ministry of Defense planned to produce 2 million FPV drones and 30,000 long-range strike drones in 2025, matching Ukraine's output and in some categories exceeding it. The Lancet loitering munition — made by ZALA Aero, a subsidiary of the Kalashnikov Group — had evolved from a niche weapon into a mass-produced staple, with improved variants featuring fiber-optic guidance that was immune to electronic jamming. Both sides had reached the same conclusion independently. The math of attrition favored the side that could produce the most expendable weapons the fastest, regardless of how many individual drones missed their targets.

The math was the same for both sides. Lose the drone, build another one by Tuesday. Lose the tank, and it is gone.

The Expected Cost of a Kill

The arithmetic of attrition warfare has a cold precision to it. Consider a single FPV drone attack against a Russian armored vehicle — a T-72B3 main battle tank, valued at approximately $2 million. The FPV drone costs roughly $400 when fitted with a warhead capable of penetrating the tank's top armor, where the protection is thinnest. The hit rate for FPV drones against armored vehicles, based on Ukrainian operational data and independent analysis, runs approximately 30% — meaning seven out of ten drones miss, malfunction, are jammed, or are shot down before reaching their target.

A 30% hit rate sounds like failure. It is not. At $400 per drone, three drones must be expended on average to achieve one kill. The expected cost per destroyed tank is $1,300 — three drones at $400 each, with the third finding its mark. The attacker has just traded $1,300 for $2 million. The cost-exchange ratio runs 1,538 to 1 in the attacker's favor.

The attacker can absorb a 70% failure rate and still dominate the cost curve. The defender cannot absorb even a single loss without compounding strategic damage, because the tank cannot be replaced on a production timeline that matches the drone's. A Ukrainian FPV workshop produces dozens of drones per day. A Russian tank factory produces a handful per month, and many of those are refurbished Soviet-era hulls pulled from storage, not new-production vehicles.

This is the core of the attrition math: when the weapon is cheap enough to be expendable, failure does not compound. When the target is expensive enough to be irreplaceable, every loss compounds. The cheap side wins not by being accurate but by being prolific. The expensive side loses not by being outfought but by being outproduced.

Attritable by Design

Military strategists have a word for platforms designed to be lost: **attritable**. The term entered the defense lexicon in the early 2020s as the Pentagon began grappling with the implications of cheap autonomous systems, but the concept is older than the word. An attritable platform is one whose loss does not create a strategic penalty. Its cost is low enough, and its production rate is fast enough, that destroying one is like pouring a cup of water from a river — the river does not notice.

The FPV drone is the most attritable weapon system in the history of warfare. At $200 to $500 per unit, with a production cycle measured in hours and a supply chain built from globally available commercial components, the FPV drone can be produced at rates that absorb any loss rate an adversary can inflict. Ukraine's 500-plus manufacturers form a distributed production network with no single point of failure — destroy one workshop, and the others continue producing. The decentralization is not a design choice. It is a survival strategy, born from the reality of operating under Russian missile and drone strikes.

Attrition-by-design extends beyond the individual drone. The entire drone warfare ecosystem in Ukraine operates on attritable principles. Operators are trained in weeks, not years. Production tooling — 3D printers, soldering stations, CNC routers — costs thousands of dollars, not millions. Software updates are pushed over encrypted channels and installed in the field. When Russia deploys a new electronic warfare system that jams a particular frequency band, Ukrainian engineers shift to a different band. When the Russians adapt to that, the Ukrainians shift again. The adaptation cycle runs in days, not the months or years that characterize traditional military procurement.

The Russian side has learned the same lesson. The Lancet program began as a precision loitering munition and has evolved into a mass-production platform with multiple variants, each adapted to counter specific Ukrainian defenses. Russian FPV production has scaled from artisanal workshops to industrial facilities, mirroring Ukraine's trajectory with a delay of roughly six to twelve months.

The Scale of Mutual Destruction

The numbers coming off the Ukrainian and Russian front lines in 2024 and 2025 defy comparison with any previous conflict. Both sides were losing an estimated 10,000 drones per month by mid-2024 — a combined destruction rate exceeding any weapons expenditure in modern war outside of artillery ammunition. Ukrainian sources reported that drones accounted for more armored vehicle kills than any other weapon system, surpassing anti-tank guided missiles, mines, and artillery combined.

The financial scale of this destruction was staggering in absolute terms but revealing in relative terms. If Ukraine expended 10,000 FPV drones per month at an average cost of $400 each, the monthly drone expenditure was $4 million. If those drones destroyed even a small fraction of their targets — say, 300 armored vehicles, trucks, and artillery pieces per month, a figure consistent with verified open-source tracking — the value of destroyed Russian equipment ran into the hundreds of millions of dollars. The cost-exchange ratio across the entire theater, aggregated over months and years, remained overwhelmingly in the attacker's favor.

This is what attrition warfare looks like in the drone age. Both sides are bleeding. Both sides are producing. Both sides are adapting. But the fundamental asymmetry holds: the drone costs hundreds of dollars; the target costs millions. No production rate on the defending side can overcome that ratio. The defender's only options are to produce enough counter-drone systems to tip the equation, to produce enough targets to absorb the losses (economically impossible at current cost ratios), or to change the nature of the battlefield so that the drone's advantages are neutralized.

No one has managed the third option yet.

The Crowdfunding War

One of the most unusual features of Ukraine's drone war is the degree to which it has been funded by civilians. The Army of Drones program, administered through Fedorov's Ministry of Digital Transformation, created a pipeline for civilian donations to flow directly into military drone procurement. Individual campaigns raised millions of dollars in hours. The Sea Baby campaign, which funded production of 35 Magura V5 maritime drones, raised $7.8 million in 36 hours. Smaller campaigns funded FPV drone units at the battalion level, with donors sometimes receiving video footage of their specific donation in action — a grim form of crowdfunding reward.

The crowdfunding model revealed something important about the economics of the $200 war. The barrier to entry for a precision strike capability had dropped below the threshold of individual philanthropy. A single donor contributing $10,000 funded fifty FPV drones. A community fundraiser organized through a Telegram channel could equip an entire drone unit. The traditional relationship between industrial capacity and military capability — where only nation-states with large economies could sustain a war — had been partially severed. Not entirely, and not permanently. But enough that a country with a GDP one-tenth the size of its adversary's could sustain a drone attrition campaign for years.

This was not how wars were supposed to work. The defense economics textbooks, written in an era of $100 million fighter jets and $13 billion aircraft carriers, assumed that the cost of waging war scaled with national wealth and industrial capacity. The drone age had introduced a different variable: the cost of waging war now scaled with the price of commercial electronics. And commercial electronics were getting cheaper every year.

The Lesson No One Wants to Learn

The attrition math emerging from Ukraine carries an implication that extends far beyond the Eastern European front. If a $400 FPV drone with a 30% hit rate yields an expected cost per kill of $1,300 against a $2 million armored vehicle, then the entire logic of armored warfare — the premise that heavily protected mobile platforms are worth their extraordinary cost because they survive on the battlefield — is under question. Not eliminated. The tank is not dead. But the economic assumptions that justified decades of investment in heavy armor, at $2 million to $10 million per vehicle, are no longer self-evident.

The same logic applies to naval vessels, to combat aircraft, and to every other platform whose value depends on the assumption that its cost is justified by its survivability. When the cost of threatening that platform drops to $200, the survivability calculation changes. Not in theory. On the battlefield, right now, in real time.

The attrition math plays out on land in Ukraine. But the most dramatic demonstration of the cost inversion is not on land at all. It is at sea — where a non-state actor in one of the world's poorest countries spent less than $2 million to impose over $1 billion in costs on the most powerful navy on Earth.

Chapter 2: The Billion-Dollar Blockade

In late 2023, Emily Harding pulled up the latest shipping data on her screen at the Center for Strategic and International Studies in Washington, D.C. Harding, a former CIA analyst and the deputy director of CSIS's International Security Program, had spent two decades tracking threats that defied conventional military categories — asymmetric warfare, cyber operations, supply chain vulnerabilities. What she saw in the Red Sea transit data was something new. In the three months since Ansar Allah — the Houthi movement controlling much of northern Yemen — had begun targeting commercial vessels with cheap drones and anti-ship missiles, Red Sea transit volume had dropped by 70%. The largest disruption to international shipping since the COVID-19 pandemic had been imposed not by a global plague or a great-power blockade, but by a rebel group in one of the world's poorest countries.

Harding cross-referenced the shipping data with US Navy missile expenditure reports. The USS *Carney* (DDG-64), an Arleigh Burke-class guided-missile destroyer, had fired more interceptors in three months than most destroyers fire in a career. The USS *Mason* (DDG-87) and USS *Laboon* (DDG-58) had added their own expenditures. Across the fleet, more than 200 interceptors had been launched against Houthi drones and missiles since November 2023. The cost of those interceptors exceeded $1 billion.

The Houthis' entire drone and missile arsenal — the weapons that had triggered more than $1 billion in American defensive spending — cost a fraction of that figure. The asymmetry was not subtle. The cost-exchange ratio was running 100 to 1 in the Houthis' favor, and it was getting worse with every engagement.

Harding's analysis, and those of her colleagues at CSIS, RAND, and the Congressional Research Service, would reach Capitol Hill within weeks. The central finding was not complicated, but it was politically devastating: the United States was losing the economic war in the Red Sea to a group that could not afford a single warship.

The Tempo of Cheap Weapons

The Houthi campaign did not begin as an exercise in economic warfare. Ansar Allah's leadership framed the attacks as solidarity with Palestinians in Gaza, launching the first strikes against commercial shipping in November 2023 in the wake of the October 7 crisis and Israel's subsequent military operations. But whatever the political motivation, the operational result was an economic masterclass in asymmetric cost imposition.

The Armed Conflict Location and Event Data Project (ACLED) tracked 417 drone-related incidents in the 12 months following October 2024 — a sevenfold increase from the 59 incidents recorded in the prior year. The Houthis were not slowing down. They were accelerating. Each wave of attacks was larger, more diverse in its targeting, and more difficult to intercept. The operational tempo — the rate at which attacks were launched — exceeded anything the US Navy had planned for in its Red Sea force posture. The Navy had deployed destroyers capable of defending themselves and nearby commercial vessels against individual threats. The Houthis were launching swarms.

A single Houthi attack wave might include a mix of one-way attack drones, anti-ship ballistic missiles, and anti-ship cruise missiles, arriving from different vectors at different altitudes at different speeds. The defenders had seconds to classify each threat, assign an interceptor, and fire. The attacking side had days or weeks to prepare the wave, at a total cost measured in hundreds of thousands of dollars. The defending side expended interceptors costing millions per engagement in minutes.

The math did not require a RAND analyst to understand. But the RAND analysis made it impossible to ignore.

The tempo also revealed a logistical asymmetry that compounded the cost problem. The Houthis could replenish their drone stocks in weeks through the Iranian supply chain — disassembled weapons shipped by sea, reassembled by trained technicians in Yemeni workshops. The US Navy's interceptor missiles required months to manufacture, and the production lines were already strained by commitments to Taiwan contingency planning, European deterrence, and the US-Israel combined operations against Iran in 2024. Every SM-6 fired at a Houthi drone was an SM-6 unavailable for the Pacific theater. The Red Sea was not just draining the Navy's budget. It was draining its magazine depth — the total number of high-end missiles available across the entire fleet — at a rate that classified Pentagon assessments reportedly described as unsustainable.

The Price of a Magazine

The interceptor cost trap is the naval expression of the same attrition math playing out on land in Ukraine, but with larger numbers and higher stakes. The US Navy's primary air defense missiles — the weapons used to shoot down Houthi drones and missiles — are among the most expensive munitions in the American arsenal.

The SM-2 Standard Missile, the workhorse of fleet air defense, costs approximately $2.1 million per round. The SM-6 Extended Range Active Missile, which provides the Navy's longest-range air defense capability, costs approximately $4 million to $5 million depending on the variant. The Evolved SeaSparrow Missile (ESSM) runs roughly $1.2 million. And the SM-3, designed for ballistic missile defense, costs between $9.6 million and $27.9 million per round depending on the variant.

RAND analysts and defense budget specialists estimated that through mid-2024, the US Navy had fired approximately 120 SM-2s, 80 SM-6s, and a smaller number of ESSMs and SM-3s in defense of commercial shipping and military vessels in the Red Sea. The total interceptor expenditure exceeded $1 billion.

The weapons those interceptors were designed to shoot down cost a fraction of the defensive response. A Shahed-pattern one-way attack drone — the type Iran has supplied to the Houthis in large quantities — costs between $50,000 and $80,000 per unit when produced domestically by Russia, and was originally sold by Iran at approximately $193,000 per unit. The more sophisticated anti-ship ballistic missiles in the Houthi arsenal cost more, but even the most expensive Houthi weapons represent a small fraction of the interceptors used to defeat them.

The arithmetic of a single engagement illustrates the trap. A Houthi drone swarm of 50 weapons — a mix of one-way attack drones and missiles — costs approximately $1.5 million to launch. The defensive response, assuming each threat requires at least one interceptor and some require two, runs approximately $150 million. The cost-exchange ratio: 100 to 1 in the Houthis' favor.

The USS *Carney*'s missile magazine holds 90 vertical launch cells. Each cell can carry one SM-2, one SM-6, or one ESSM quad-pack (four missiles per cell). The magazine is the ship's lifeblood — when it is empty, the destroyer must leave station and sail to a rearming facility, a transit measured in days. The Houthis do not need to sink a destroyer. They need only to empty its magazine. At the rate of expenditure observed in 2023-2024, that process took weeks, not months.

The Supply Chain of Disruption

The Houthi drone and missile arsenal was not indigenous. The weapons came overwhelmingly from Iran, which had spent years building a drone export capability that now supplied proxy forces across the Middle East — Hezbollah in Lebanon, allied militias in Iraq, and Ansar Allah in Yemen. The Shahed-series drones were manufactured in Iran, disassembled for shipping, and reassembled by Houthi technicians trained by Iranian advisors. The supply chain was robust, difficult to interdict, and remarkably cheap.

But the Iranian connection was only part of the story. In February 2025, the US Navy intercepted a shipment bound for Houthi-controlled territory that contained, among other items, jet engine components of Chinese manufacture. The components, analysis revealed, could be used to produce drone engines approximately four times faster than the designs the Houthis had previously employed. The supply chain feeding the Houthi drone campaign was not a simple Iran-to-Yemen pipeline. It was a network that touched Chinese manufacturers, Iranian assembly plants, maritime smuggling routes, and end users who could integrate commercial and military components with increasing sophistication.

The dual-use nature of the components made interdiction extraordinarily difficult. A jet engine part that could power a drone could also power a generator. A flight controller that could guide a weapon could also stabilize a survey drone. The same export control challenge that plagued efforts to restrict the FPV drone supply chain in Ukraine applied at sea. The components of precision strike were indistinguishable from the components of civilian commerce.

The Trade War No One Declared

The military cost of the Houthi campaign was significant. The economic cost was worse.

The 70% drop in Red Sea transit volume did not mean ships stopped moving goods. It meant they took the long way around — routing south around the Cape of Good Hope, adding 10 to 14 days and hundreds of thousands of dollars in fuel costs to each voyage. The shipping industry, which moves approximately 80% of global trade by volume, absorbed the cost and passed it forward. Insurance premiums for Red Sea transit surged. Freight rates spiked. Delivery timelines stretched. The downstream effects rippled through global supply chains, from European manufacturing to Asian exports to American consumer prices.

CSIS estimated the total economic impact of the Houthi campaign at tens of billions of dollars in additional shipping costs, insurance premiums, and supply chain disruptions through mid-2024. The US military response — Operation Prosperity Guardian, the multinational naval task force assembled to protect commercial shipping — cost billions more in operational expenses, interceptor expenditures, and force deployment.

The Houthis' total weapons expenditure for the entire campaign, across all drones and missiles launched, was estimated at less than $100 million.

The cost-exchange ratio, calculated across the full scope of military and economic impacts, was not 100 to 1. It was closer to 1,000 to 1.

A group that could not afford a single warship, operating from one of the poorest countries on Earth, had imposed costs on the global economy that dwarfed the total GDP of Yemen. The Houthi campaign was not a military operation in any traditional sense. It was an exercise in economic leverage executed with weapons that cost less than the fuel it took to dodge them.

The Strategic Verdict

The Red Sea campaign proved something that the Ukraine conflict had suggested but not yet confirmed: the cost inversion extends beyond the battlefield to global trade, alliance credibility, and great-power deterrence.

In Ukraine, cheap drones destroyed expensive military hardware. The cost was borne by militaries. In the Red Sea, cheap drones disrupted expensive shipping routes. The cost was borne by the global economy. The second application is more dangerous, because the economic damage is not bounded by the theater of conflict. A drone launched from Yemen can impose costs on a container ship bound for Rotterdam, on an insurance company in London, on a consumer in Ohio.

The strategic implication was not lost on military planners in Beijing, Tehran, Moscow, or any other capital whose strategists were watching the Red Sea unfold. The Houthi model demonstrated that a non-state actor with access to cheap drones and a chokepoint could hold global trade hostage at a cost-exchange ratio that no conventional military response could overcome. The US Navy could shoot down every drone the Houthis launched and still lose the economic war, because every interceptor cost more than every drone, and every diverted ship cost more than every interceptor.

The Houthis built their drone air force from Iranian designs and Chinese components. Ukraine built its from DJI parts and 3D printers. Both proved the same thing: the weapons reshaping warfare are not being designed in government laboratories. They are being assembled from the global consumer electronics supply chain — and the center of that supply chain is a single city in southern China.

Chapter 3: Made in Shenzhen

On a Tuesday morning in the spring of 2025, a quality control manager named Chen Wei walked the floor of an electronics assembly plant in Shenzhen's Bao'an district. Chen had worked in the Bao'an manufacturing zone for fourteen years, first in mobile phone assembly, then in the booming consumer drone components sector. The plant where he now supervised production occupied 12,000 square meters of floor space and employed 340 workers across two shifts. Its automated assembly lines produced brushless motors, electronic speed controllers, lithium-polymer battery packs, flight controllers, and CMOS camera sensors — the fundamental building blocks of every multirotor drone on Earth.

The daily output filled shipping containers bound for six continents. The largest orders went to DJI's distribution network — the Netherlands, the United States, Japan. Midsize orders went to Amazon fulfillment centers in Kentucky and wholesale electronics bazaars in Dubai. Smaller orders, placed through intermediary trading companies whose end customers were obscured by layers of commercial invoicing, went to destinations that Chen did not track and his employer did not advertise. The plant's sales team dealt in purchase orders and wire transfers, not geopolitics.

But geopolitics dealt in their products. The same $3 brushless motor that powered a toy quadcopter sold on Amazon powered the FPV drone that destroyed the Tor missile system near Kupiansk. The same $8 flight controller that stabilized a DJI Mini above a wedding reception in California stabilized the weapon that sank a Russian patrol vessel in the Black Sea. The same $2 electronic speed controller that regulated the rotor speed of a child's racing drone regulated the approach velocity of a Houthi one-way attack drone bearing down on a commercial tanker in the Red Sea.

Chen Wei did not build weapons. He built components. The distinction, in the age of the $200 war, had become meaningless.

The DJI Factor

No company better illustrates the dual-use paradox than Da-Jiang Innovations — DJI — the Shenzhen-headquartered company that dominates the global consumer drone market with over 70% market share. DJI does not manufacture military drones. Its products are consumer and commercial platforms — photography drones, agricultural sprayers, survey tools, inspection systems. Its founding engineer, Frank Wang, started the company in a university dorm room in 2006 with a focus on flight controller technology. By 2024, DJI had grown into a global brand with revenues exceeding $4 billion and products sold in more than 100 countries.

DJI's relevance to the $200 war is not about the company's intentions. It is about its supply chain. DJI pioneered the miniaturization and cost reduction of drone components — brushless motors, inertial measurement units, flight controllers, gimbal stabilization systems, high-resolution cameras — that made consumer drones affordable at scale. In doing so, DJI created the component ecosystem from which the world's cheapest weapons are assembled.

Ukrainian FPV drone workshops do not buy DJI drones and strap warheads to them (though that has happened). They buy DJI-standard components — motors, electronic speed controllers, camera modules — from the same Shenzhen suppliers that feed DJI's own assembly lines, or from DJI's authorized parts distributors, or from the vast aftermarket of compatible components produced by dozens of Chinese manufacturers who reverse-engineered DJI's specifications. The DJI component standard has become, inadvertently, the de facto standard for FPV drone warfare. When a Ukrainian engineer designs a new FPV drone frame, the motor mounts are spaced for DJI-compatible motors. When a Russian workshop builds a Lancet variant, the flight controller speaks the same protocols. The supply chain does not distinguish between customer and combatant.

The US government has recognized the problem. In 2024, the Federal Communications Commission (FCC) placed DJI on its Covered List, restricting the use of DJI communications equipment in US networks. Congress considered legislation to ban DJI drones from US government use. In April 2025, the United States enacted a 170% import duty on Chinese drones and drone components — the most aggressive trade measure ever applied to the drone sector.

The tariff was a signal. It was not a solution.

The Tariff Wall and Its Limits

The 170% duty on Chinese drone imports represented the US government's most direct attempt to sever the dependency between the American drone market and the Shenzhen supply chain. The intent was twofold: punish China for its role in the drone proliferation pipeline, and incentivize domestic production of drone components that could replace Chinese suppliers.

The problem was structural. The Chinese drone component industry had spent two decades building the manufacturing infrastructure, supply chain relationships, and production expertise that made it the cheapest and most reliable source of drone parts on the planet. Alternative sources existed — manufacturers in Taiwan, South Korea, Vietnam, and increasingly in the United States and Europe — but none operated at China's scale, and none could match China's prices. A brushless motor produced in Shenzhen for $3 cost $8 to $12 when sourced from a non-Chinese manufacturer, and delivery timelines were measured in months rather than weeks.

The tariff wall would, over time, encourage investment in alternative supply chains. American companies like Skydio were already building domestic drone manufacturing capacity — Skydio's facility in San Mateo, California, assembled its X10D reconnaissance drones from largely non-Chinese components. But Skydio's drones cost tens of thousands of dollars each, not hundreds. The economics of a $200 FPV drone depended on Shenzhen pricing. Remove Shenzhen from the equation, and the FPV drone cost $600 or $800 — still cheap enough to be attritable, but not cheap enough to sustain the 120,000-to-1 cost-exchange ratios that defined the revolution.

More fundamentally, the tariff addressed the American market. It did nothing about the global supply chain. A Ukrainian drone manufacturer buying Shenzhen motors through a European intermediary was unaffected by US tariffs. A Houthi weapons assembler receiving components via Iran was unaffected. The global flow of dual-use electronics from China to every corner of the planet continued unimpeded by any single nation's trade policy.

The drone revolution was not being built on a supply chain that could be sanctioned away. It was being built on the same supply chain that produced every smartphone, every laptop, every consumer electronic device on Earth. To embargo the $3 brushless motor was to embargo the global consumer economy.

The Iran Pipeline

If DJI represented the inadvertent side of the Shenzhen-to-battlefield pipeline, Iran represented the deliberate side. Tehran had spent more than a decade building a drone program that drew on Chinese commercial components, Iranian engineering, and a global procurement network designed to evade sanctions.

The Shahed series — the one-way attack drones that Russia has launched by the thousands against Ukrainian cities and that Houthi forces have fired at commercial shipping in the Red Sea — illustrates the pipeline. The Shahed-136 is manufactured at facilities in Iran, primarily at the Iran Aircraft Manufacturing Industrial Company (HESA) complex in Isfahan. Its airframe is Iranian-designed. Its engine is derived from a commercially available design. Its guidance system uses components available on the global market. Iran originally sold the Shahed-136 to Russia at approximately $193,000 per unit; Russian domestic production has since brought costs down to an estimated $70,000 to $80,000 per unit — still a fraction of the cost of any comparable Western munition.

In early 2026, defense media reported that China had transferred approximately 1,000 loitering munitions to Iran through Pakistan under an arrangement linked to oil-for-arms trading. The transfer represented a significant escalation in China's willingness to supply military drone systems — not just components, but complete weapons — to a state that functions as the primary drone supplier to proxy forces across the Middle East.

The China-Iran drone pipeline fed a network that extended to the Houthis in Yemen, Hezbollah in Lebanon, and allied militia forces in Iraq and Syria. Iran supplied not only complete weapons systems but also the technical training, manufacturing know-how, and component supply chains that allowed its proxies to establish indigenous production capabilities. The Houthis were no longer merely receiving and launching Iranian drones. They were assembling and modifying them, adapting designs to counter specific defensive systems, and building an operational capability that would persist even if Iranian supply were cut.

The components in that pipeline traced back, through layers of intermediaries, to the same Shenzhen manufacturing ecosystem that supplied DJI, Amazon, and the global consumer electronics market. A flight controller chip fabricated in a Shenzhen foundry might be sold to an Iranian trading company in Dubai, shipped to a port in Bandar Abbas, trucked to an assembly facility in Isfahan, integrated into a Shahed airframe, shipped disassembled to Yemen, reassembled by Houthi technicians, and launched at a commercial tanker in the Red Sea — all within a few months, through a chain of transactions that no single export control regime could monitor end to end.

The Component Archaeology of Conflict

The forensic evidence confirmed what the economics predicted. When international investigators examined the debris of Houthi drones recovered from the Red Sea, and when Ukrainian bomb disposal teams disassembled unexploded Russian Shaheds that fell in Ukrainian fields, they found the same thing: commercial components manufactured by companies in China, the United States, Europe, and Japan, routed through intermediaries in the

Gulf states and Central Asia.

A single Shahed-136 recovered in Ukraine contained, according to analysis by the Conflict Armament Research group and independent investigators, components from more than a dozen countries. Microprocessors from US-designed chip architectures fabricated in Asian foundries. Servo motors from Japanese manufacturers. GPS modules from European suppliers. And the basic structural components — motors, speed controllers, batteries, wiring — from Shenzhen.

The component archaeology of the Shahed illustrated the fundamental impossibility of controlling drone proliferation through traditional export controls. The Missile Technology Control Regime (MTCR), established in 1987 to limit the spread of missile technology, was designed for an era when precision strike required identifiable, purpose-built military hardware — complete missile systems, specialized guidance packages, military-grade propulsion units. The MTCR could restrict the sale of a cruise missile. It could not restrict the sale of a $3 brushless motor, a $2 electronic speed controller, and a $8 flight controller, because those components had legitimate civilian applications that dwarfed their military use.

The Wassenaar Arrangement on Export Controls for Conventional Arms and Dual-Use Goods and Technologies faced the same structural limitation. Wassenaar could restrict the export of a complete military drone system. It could not restrict the export of the components from which a military drone system could be assembled, because those components were indistinguishable from the components of civilian commerce.

The structural irony was complete. The same consumer electronics revolution that put a smartphone in every pocket and a drone on every Christmas list had also put the components of precision strike in every hardware market on the planet. The revolution could not be reversed without reversing the globalization of electronics manufacturing itself. And no government — not the United States, not the European Union, not China — was willing to do that.

The Unavoidable Conclusion

The Shenzhen supply chain is the foundation of the $200 war. Every FPV drone in Ukraine, every Shahed in Iran's arsenal, every Houthi weapon in the Red Sea, and every hobbyist quadcopter on Amazon share the same manufacturing DNA. They are assembled from the same components, produced in the same factories, shipped through the same logistics networks. The distinction between consumer and combatant, between toy and weapon, between civilian technology and military capability, exists on paper. It does not exist on the factory floor.

This fact has an implication that no tariff, no embargo, and no export control regime can solve: the drone revolution is inseparable from the global consumer electronics supply chain. You cannot embargo the $3 brushless motor without shutting down Amazon. You cannot restrict the $8 flight controller without grounding every agricultural drone, every film crew drone, every search-and-rescue drone on the planet. The weapon *is* the supply chain.

The Shenzhen supply chain is not new. What is new is the money pouring into it — not from governments, but from Silicon Valley venture capitalists who have decided that the economics of cheap autonomous weapons represent the investment opportunity of a generation.

Chapter 4: The $49 Billion Bet

At 9:30 a.m. on March 17, 2026, the ticker symbol SWMR appeared on the Nasdaq display at MarketSite in Times Square. The screens above the trading floor, five stories tall and visible from the intersection of Broadway and 43rd Street, showed a stock price of $5.00. Swarmer — a Ukrainian drone-swarm software company born in the first months of Russia's full-scale invasion — had completed its initial public offering.

Swarmer's credentials were unlike anything Wall Street had seen before. The company had logged more than 100,000 combat missions, not in simulation but in active warfare against the Russian military along a 600-mile front. Its core product, a software platform called Styx, allowed a single operator to manage a swarm of up to 25 drones simultaneously, coordinating autonomous targeting, navigation, and strike sequencing across multiple platforms. Styx had been tested under electronic warfare conditions that would have grounded most commercial autonomy software. It had been updated in the field, sometimes daily, by engineers working from reinforced basements within range of Russian artillery. Its reliability data came not from a controlled test range in the California desert but from the most heavily contested electromagnetic environment on the planet.

Within two trading sessions, SWMR surged from $5.00 to $65.04 — a gain of approximately 1,200%.

The market was not buying a startup. It was buying combat-proven autonomous warfare software. And it was paying a thousand-percent premium for the privilege.

The Flood

Swarmer's IPO was not an isolated event. It was the most visible expression of a financial transformation that had been accelerating for two years. In 2025, venture capital investors poured $49.1 billion into defense technology companies — nearly double the $27.2 billion invested in 2024. The number of VC firms actively investing in defense technology increased by 41% year over year. Mainstream venture funds that had previously avoided defense — citing ethical concerns, reputational risk, or the perceived incompatibility of Silicon Valley culture with military procurement — dropped their objections and entered the sector.

The reframing was striking in its speed and its coordination. Defense technology, which as recently as 2018 had triggered employee walkouts at Google over Project Maven, was now marketed as "supporting democratic values" and "protecting the rules-based international order." The language changed. The check sizes did not shrink. In boardrooms along Sand Hill Road and in converted warehouses in Manhattan's Flatiron district, partners who had once refused to take meetings with defense founders were now competing for allocation in oversubscribed rounds.

The money followed a thesis that could be stated in a single sentence. Cheap autonomous systems will permanently replace expensive legacy platforms, and the company that builds the winning platform will capture the largest addressable military market in history.

The Valuation Map

The capital flowing into defense technology was not evenly distributed. It concentrated in a handful of companies whose valuations reflected the market's conviction that the cost inversion was permanent and that autonomy was the winning strategy.

Anduril Industries, founded by Palmer Luckey in 2017, reached a valuation of $60 billion in its March 2026 funding round — the highest private valuation in defense-technology history. Anduril's Lattice software platform, which integrates sensor data across autonomous systems into a unified command-and-control interface, had become the connective tissue for a growing portfolio of autonomous weapons platforms. In March 2026, Anduril won a US Army contract with a ceiling value of $20 billion — the largest single contract ever awarded to a defense-technology startup.

Behind Anduril, a cluster of companies had reached valuations that would have been unthinkable five years earlier. Chaos Industries, focused on autonomous vehicle systems for defense, reached $4.5 billion. Saronic Technologies, building autonomous naval vessels, reached $4 billion. Shield AI, whose Hivemind autonomous piloting software flew the V-BAT vertical takeoff drone without GPS or communications, held a valuation of $2.8 billion as of its November 2023 Series F round. In Europe, Helsing — a defense AI company based in Munich — reached a valuation of 12 billion euros, making it the most valuable defense-technology startup on the continent.

Each of these companies was built on the same fundamental premise: that the future of military capability belonged to autonomous software platforms running on cheap, mass-produced hardware, not to exquisite manned platforms costing hundreds of millions of dollars each. The VCs were not betting on a technology. They were betting on an economic shift.

The Unit Economics of War

The venture capital thesis had a quantitative backbone. Traditional defense platforms occupy the top of the cost curve — a domain that the VC community had analyzed with the same rigor it applied to software-as-a-service unit economics or marketplace take rates.

An F-35A Lightning II costs approximately $80 million per airframe, with lifetime operating costs exceeding $400 million per aircraft over a 30-year service life. A Virginia-class nuclear attack submarine costs approximately $4.3 billion. A Gerald R. Ford-class aircraft carrier costs approximately $13 billion, not including its air wing. These platforms represent the most expensive expression of American military power — and they are the platforms most vulnerable to the cost inversion.

The VC thesis held that attritable autonomous systems costing between $500 and $50,000 each could perform an increasing share of the missions currently assigned to these platforms, at a fraction of the cost and without risking human operators. A $50,000 loitering munition could suppress an air defense site that previously required an $80 million manned aircraft to attack. A $500 FPV drone could destroy an armored vehicle that a $4 million Javelin missile was designed to kill. A $250,000 autonomous sea drone could threaten a naval vessel that a $2 billion submarine was designed to sink.

The unit economics were not theoretical. They had been demonstrated in combat. Ukrainian FPV drones had destroyed more than a thousand armored vehicles. Magura V5 sea drones had struck Russian warships. Swarmer's Styx system had coordinated autonomous multi-drone strikes in active warfare. Every engagement produced data that reinforced the thesis: cheap, autonomous, expendable systems were not merely cost-effective alternatives to legacy platforms. They were cost-effective replacements.

The VCs read the data the same way they read a SaaS company's revenue growth — as proof of product-market fit. The product was autonomous warfare software. The market was every military on Earth.

The AI-in-Drones Market

Behind the headline venture investments, a broader market was forming. Research firms tracking the defense technology sector estimated that the AI-in-drones market — encompassing autonomous navigation, computer vision targeting, swarm coordination, and AI-enabled electronic warfare — reached $845 million in 2025. The projected trajectory was steep — $10.9 billion by 2035, representing a compound annual growth rate (CAGR) of 29.8%.

That growth rate reflected not just military demand but the convergence of military and commercial AI development. The same computer vision algorithms that allowed a Skydio drone to navigate autonomously through a forest canopy could, with modification, allow an FPV drone to track a moving vehicle through its terminal approach. The same reinforcement learning techniques that trained a warehouse robot to sort packages could train a swarm of drones to coordinate an attack pattern. The same neural network architectures that powered autonomous driving could power autonomous targeting.

The convergence was a feature, not a bug, of the VC thesis. Defense-technology startups that built on commercial AI foundations could iterate faster and cheaper than traditional defense contractors who built proprietary systems from scratch. Anduril's Lattice used commercial cloud infrastructure. Shield AI's Hivemind ran on commercially available processors. Swarmer's Styx was built with open-source machine learning frameworks. The entire defense-tech startup ecosystem was, in essence, applying Silicon Valley's commercial software development model to the problem of autonomous weapons.

The implication for the cost curve was profound. When autonomous warfare software was built on commercial foundations, its development cost was amortized across both civilian and military applications. When it ran on commercial hardware, its unit cost tracked the consumer electronics curve — downward, relentlessly, year after year. The $200 drone was already cheap. The software that would make it autonomous was getting cheaper.

The Ethical Pivot

The financial rush into defense technology did not occur without friction. As recently as 2018, Google employees had staged walkouts over Project Maven — the DoD initiative to apply machine learning to drone surveillance footage analysis. More than 3,000 Google employees signed a letter demanding the company withdraw from the contract. Google did. The incident became a defining moment in the relationship between Silicon Valley and the Pentagon, establishing an apparent red line. Technology companies would not build tools for war.

The red line lasted approximately four years. By 2022, the Ukraine invasion had reframed the debate. Russian missiles striking apartment buildings in Kyiv, Russian forces occupying Ukrainian cities, and Ukrainian defenders using consumer technology to fight back created a moral narrative that Silicon Valley found compelling. Defense technology was no longer about building tools for abstract future wars. It was about helping a democratic nation defend itself against an authoritarian invader.

The reframing was genuine for many in the industry, and cynical for others, and for most it was some combination of both. The financial incentives were enormous. The $49.1 billion in 2025 VC investment represented a market opportunity that no fund manager could ignore without explaining to limited partners why the fund had sat out the largest capital reallocation in defense technology history. The ethical objections did not disappear. They were outweighed — by patriotism for some, by profit for others, by the conviction that the technology would be built regardless and it was better built by democracies than by autocracies.

Palmer Luckey, Anduril's founder, articulated the position bluntly in public appearances throughout 2025 and 2026: the people who refuse to build defense technology do not prevent the technology from being built. They merely ensure it is built by someone else — someone who may not share their values. It was an argument that had the virtue of being logically sound and the limitation of proving everything and nothing. But it was effective. The money followed.

The Swarmer Signal

Swarmer's IPO represented the culmination of a sequence that began in a Ukrainian basement in 2022 and ended on the Nasdaq trading floor in 2026. The company's trajectory compressed the entire arc of the defense-technology revolution into four years — founded in wartime, developed under fire, proven in combat, funded by venture capital, and valued by public markets.

The 1,200% surge in Swarmer's stock price in its first two trading sessions was, by any conventional measure, a speculative mania. The company had limited revenue, no established customer base outside of Ukraine, and an uncertain path to profitability. But the market was not pricing Swarmer's current financial position. It was pricing the future of autonomous warfare — a future in which combat-proven swarm software, integrated into mass-produced attritable platforms, would become the standard weapon system for every military willing to accept the cost inversion.

The Swarmer IPO sent a message that defense budget analysts, military planners, and venture capitalists all understood in different ways. Wall Street's verdict was simple: the era of the $100 million fighter jet is ending. The era of the $500 drone is beginning. $49 billion says so.

The Money's Conclusion

The venture capital flood into defense technology is the financial market's judgment on the cost inversion documented in the first three chapters of this book. The attrition math in Ukraine proved that cheap drones destroy expensive platforms. The Red Sea campaign proved that cheap drones impose billion-dollar costs on great powers. The Shenzhen supply chain proved that the components are globally available and impossible to restrict. The VC money is the market's conclusion: the cost inversion is real, it is permanent, and the companies that build the winning autonomous platforms will capture a market measured in hundreds of billions of dollars.

That conclusion carries an uncomfortable corollary. If the smart money is flowing away from expensive legacy platforms and toward cheap autonomous mass, then the institutions that still spend the overwhelming majority of their budgets on expensive legacy platforms — the US Department of Defense chief among them — are on the wrong side of the cost curve. The VC community has priced in the inversion. The Pentagon's budget has not.

The VC money says the cost inversion is real. The Shenzhen supply chain says it is unstoppable. The Red Sea campaign says it works. But is this actually new? History suggests the answer is no — and that every time the expensive side has faced a cost inversion, it has lost. The pattern starts 600 years ago, in a muddy field in northern France.

Chapter 5: The Knight and the Longbow

The rain had not stopped for days when Sir Guillaume Martel, appointed bearer of the Oriflamme — the sacred war banner of France, though historians debate whether the banner was actually carried onto the field that day — sat mounted at the rear of the French vanguard on the morning of October 25, 1415, watching the English line across a freshly plowed field near the village of Azincourt in northern France. He was approximately forty-five years old, a veteran of a dozen engagements, encased in plate armor that had cost his household the equivalent of a prosperous farm. His warhorse, bred for war and trained over years, was worth a small estate. The lance in his hand had been crafted by a specialist armorer. The sword at his hip was Toledo steel, balanced to his grip. Everything about Guillaume Martel — his equipment, his training, his lineage, his place in the formation — represented the pinnacle of six centuries of European military investment in a single weapons platform: the mounted knight.

Across the mudfield, approximately 750 yards away, stood roughly 5,000 English and Welsh longbowmen. They were yeoman farmers, recruited from the counties of England and the hills of Wales. Their bows were made of yew, cut and shaped by village bowyers. Each bow cost a few shillings — less than a knight spent on a single gauntlet. The arrows were mass-produced, fletched by the thousands in workshops across southern England. An archer could be trained to battlefield competence in months. A knight required a lifetime.

King Henry V of England knew the arithmetic. His army was outnumbered, exhausted after a 260-mile march from Harfleur, and riddled with dysentery. He had perhaps 1,500 men-at-arms and those 5,000 archers against a French force that historian Anne Curry estimates at 12,000 or more, including 8,000 to 10,000 armored men-at-arms and mounted knights — the most concentrated assembly of aristocratic military wealth in a generation.

Henry ordered his archers to drive sharpened stakes into the ground before their positions, angled forward at chest height to a horse. Then he waited. The French, packed into a front no wider than 750 yards between two dense woods, began their advance through the freshly plowed, rain-soaked field. They advanced on foot, many of them dismounted because the terrain made cavalry charges suicidal, sinking ankle-deep into the mud with every step, their plate armor weighing 50 to 60 pounds per man.

The longbowmen drew and loosed.

Fifty Thousand Arrows a Minute

The numbers that followed are the earliest large-scale demonstration of a principle that would repeat across six centuries of military history. Cheap mass annihilates expensive quality when the conditions are right.

Each English longbow could loose 10 to 12 arrows per minute. With 5,000 archers on the field, the rate of fire approached 50,000 arrows per minute — a volume of projectiles that no armor, no discipline, and no courage could overcome. The arrows did not need to penetrate plate armor to be lethal. They struck visors, found gaps at joints and armpits, killed horses, and — most critically —

created chaos in a formation already struggling through knee-deep mud. The French vanguard compressed as it funneled between the woods of Tramecourt and Azincourt. Men fell. Those behind stumbled over the fallen. The formation lost coherence. Knights in full plate, unable to rise once toppled in the mud, drowned in inches of water or were crushed by the weight of men pressing from behind.

The killing lasted less than three hours. When it ended, between 5,000 and 7,000 French soldiers lay dead, including Guillaume Martel, still clutching the Oriflamme. English dead numbered fewer than 200. Juliet Barker, in her definitive account of the battle, notes that the French losses included three dukes, at least eight counts, a viscount, and an archbishop — a catastrophic erasure of the French aristocratic military class in a single afternoon. The Constable of France, Charles d'Albret, the nominal commander of the French army, was among the dead. So were the Dukes of Alencon and Bar. The cream of French military aristocracy lay in the mud, killed by men who could not have afforded a single piece of the armor they had punctured.

The most expensive weapon system in Europe had been annihilated by the cheapest.

The Economics of the Armored Knight

To understand what died at Agincourt, consider what a mounted knight actually cost. The armor alone represented an extraordinary capital investment. A full harness of plate armor in the early fifteenth century — custom-fitted by a master armorer, articulated at every joint with riveted lames that allowed movement while maintaining protection, polished to deflect glancing blows, sometimes etched or gilded for display — cost the equivalent of what historian Clifford Rogers estimates at several years of a skilled laborer's wages, comparable in relative terms to a modern luxury automobile. The finest suits, produced by master armorers in Milan or Augsburg, could cost significantly more. The Milanese workshops of the Missaglia family and the Augsburg workshops of the Helmschmid family produced armor that was as much precision engineering as metalwork — each plate shaped to distribute the force of a lance strike across the body, each joint designed to prevent a sword point from finding the gap between plates.

A warhorse trained for battle — a destrier, bred for size and temperament, conditioned over years not to shy from noise and impact — cost as much as the armor. These animals were not simply large horses. They were purpose-bred war machines, trained to bite, kick, and trample on command, steady under the crash of lance on shield. Andrew Ayton, in his study of English military service under Edward III, documented warhorse valuations that exceeded the annual income of minor knights — a single horse worth more than its rider earned in a year.

The knight's training began in childhood and continued for a decade or more, through service as a page learning courtesy and horsemanship, then as a squire learning arms and tactics, before the accolade of knighthood. Add the cost of weapons, a squire, spare horses, feed, lodging, and transport, and a single mounted knight represented a total systems cost that dwarfed anything else on the medieval battlefield.

Rogers, whose work on the military revolutions of the Hundred Years' War remains the standard reference, calculated that the cost differential between fielding a knight and fielding an archer was not merely large — it was structural. A knight cost roughly 20 to 50 times what an archer cost to equip, train, and sustain in the field. The knight was, in the language of modern defense economics, a legacy platform: exquisitely capable in the right conditions, catastrophically expensive relative to the emerging alternative.

The Longbow as Production System

The longbow, by contrast, was a masterpiece of economic efficiency. The raw material — a stave of yew, ideally from the heartwood-sapwood boundary where compression and tension properties naturally aligned — cost almost nothing. England imported yew staves in bulk from Spain, Italy, and the Baltic, and every merchant ship entering an English port was required by statute to carry a specified number of bow staves as part of its cargo. The supply chain for longbow production was, in modern terms, a distributed manufacturing network with standardized inputs and mandated supply.

English statute law required every able-bodied man to practice archery regularly, creating a deep reserve of trained archers without dedicated military academies. The Assize of Arms of 1252, updated and reinforced by subsequent statutes, made archery practice a legal obligation. Every village had its butts — the targets where men shot on Sundays after church. The effect was a national training program that cost the crown almost nothing. The skill was maintained by the population as a civic duty, and when the crown needed archers for a campaign, the trained reserve was already there.

An archer's equipment — bow, arrows, a padded jack or brigandine for light protection, a short sword or buckler — could be supplied for a fraction of a knight's gauntlet. The arrows were the critical consumable, and England had industrialized their production. Anne Curry's research into the logistical records of Henry V's campaign reveals that the English crown ordered hundreds of thousands of arrows before the Agincourt expedition, produced in bulk by fletchers working to standardized specifications. The Tower of London's armory records show orders for arrow shafts in lots of 10,000 and 20,000, with quality specifications for shaft straightness, fletching adhesion, and arrowhead type. This was not cottage industry. This was mass production, six centuries before the term was coined. The longbow was not merely a weapon — it was a production system, cheap to manufacture, quick to train on, and lethal at scale.

The Refusal to Adapt

Agincourt was not the first time the longbow had shattered a mounted charge. At Crecy in 1346 — sixty-nine years earlier — English longbowmen had destroyed a French army under almost identical conditions: a massed advance of armored knights into concentrated archery fire, with terrain funneling the attackers into a killing zone. At Poitiers in 1356, the pattern repeated. At Nicopolis in 1396, French knights charging Ottoman positions met the same concentrated missile fire with the same catastrophic results. The longbow's dominance over the mounted knight was not a single surprise but a demonstrated, repeated, empirically validated fact across seven decades of European warfare.

The French knew this. They had been on the receiving end of it for three generations. Their own commanders had written about it. Their own councils had debated it. And yet at Agincourt, the French battle plan called for an advance directly into massed longbow fire across an open, muddy field — the single worst tactical environment for heavy armor against ranged weapons.

The institutional explanation matters more than the tactical one. The mounted knight was not merely a weapons platform. He was a social institution. Knighthood was hereditary. The aristocratic families that supplied the knights also supplied the commanders, the counselors, and the political leadership of France. To admit that the knight was obsolete was to admit that the social class that defined itself through knightly combat was obsolete. The military technology and the social structure were inseparable. As Rogers observed, the French did not fail to recognize the longbow's effectiveness — they failed to accept what that effectiveness meant for their entire way of organizing military power.

Maurice Keen, in his seminal study of chivalry, documented how deeply embedded the knightly ideal was in the French aristocratic psyche. To be a knight was not a job description. It was an identity. The knight fought on horseback because that was what a knight did. The knight charged because charging was the knightly act. To suggest that the knight should dismount, disperse, or — worst of all — yield the decisive role to commoner archers was to suggest that the aristocratic warrior class should surrender its defining function. The institution would rather lose battles than lose its identity.

This is the part of the pattern that matters most for the argument of this book. The French did not lack intelligence. They lacked institutional willingness to act on what the intelligence showed them. The evidence was unambiguous — cheap, massed ranged firepower killed expensive, individual armored platforms — and the response was to send the expensive platforms forward again, because the institution could not imagine itself without them.

The Cost Inversion, Stated Plainly

Strip away the mud, the heraldry, and the six centuries of distance, and Agincourt reduces to a cost-exchange ratio. The English spent almost nothing per unit of lethality. Each arrow cost pennies. Each archer cost a fraction of a knight. The rate of fire multiplied the investment: 5,000 archers loosing 50,000 arrows per minute meant that the total English expenditure per minute of engagement was trivially small relative to the value of the targets destroyed.

The French spent enormously per unit of combat power. Each knight represented decades of investment. Each horse represented years of breeding and training. The total French expenditure on the force that crossed that mudfield — the accumulated capital of the French aristocratic military establishment — was destroyed in hours by a force that cost a small fraction of what it killed.

Consider the economics from the French crown's perspective. The dead at Agincourt included three dukes, eight counts, the Constable of France, and more than a hundred knights of the highest rank — men whose military equipment alone represented a capital stock equivalent to several prosperous estates. The ransom value of the prisoners Henry took, and the political vacuum left by the dead, destabilized French governance for a generation. The English had destroyed not just an army but an investment portfolio — decades of accumulated military capital, erased in an afternoon by a commodity that cost pennies per unit. The knight was the main battle tank of 1415. The longbow was the FPV drone. Six hundred years later, the math has not changed.

What Agincourt Teaches the Pentagon

The parallel is not metaphorical — it is structural. The US military's most expensive platforms — the F-35A Lightning II at approximately $170 million per aircraft when development costs are included, the Gerald R. Ford-class carrier at $13 billion, the Columbia-class submarine at $9 billion — represent the same category of investment as the mounted knight. They are exquisitely capable. They are the product of decades of development. They embody the institutional identity of the services that operate them. The Air Force defines itself through air superiority fighters. The Navy defines itself through carrier strike groups. The Army defines itself through armored maneuver. To suggest that a $200 drone might render these platforms strategically obsolete is to suggest that the institutional identities built around them must change.

The institutional attachment runs deeper than budget lines. Promotion boards favor officers who have commanded expensive platforms. Congressional delegations protect the factories that build them. Defense contractors have organized their workforce, their supply chains, and their shareholder expectations around multi-decade production runs of complex, costly systems. The F-35 program alone sustains an estimated 254,000 jobs across 45 states, according to Lockheed Martin — a political fact that is, in its own way, as powerful as the military fact of the longbow's lethality. The knight persisted because the feudal system needed him. The legacy platform persists because the defense-industrial system needs it.

The French aristocracy faced the suggestion that their defining weapons platform was obsolete at Crecy, at Poitiers, and at Agincourt. They heard it three times across seven decades. They did not adapt.

The question is not whether the cost inversion between cheap drones and expensive platforms is real. The evidence presented in the first four chapters of this book — from the Red Sea to Ukraine to the Shenzhen supply chain to the venture capital markets — establishes that it is. The question is whether the institution that has built its identity around the expensive platform can accept the implications before the battlefield forces the lesson, as it forced it on the French at Agincourt.

The history of that mudfield in northern France suggests the answer. Institutions do not voluntarily abandon the platforms around which they have organized their power, their budgets, their careers, and their identities. They ride them into the mud.

Agincourt was fought on land. But the cost inversion has destroyed military assumptions at sea, too — and the naval version of this story produced a response that looks eerily familiar to anyone watching the counter-drone industry today. The expensive side did not ignore the cheap threat. It built an expensive counter-system. And the cycle began again.

Chapter 6: The Torpedo and the Carrier

Commander Eli Reich had been tracking the Japanese formation for six hours.

It was November 21, 1944, and the submarine USS *Sealion* (SS-315) was running on the surface in heavy seas northwest of Formosa, her radar picking up a cluster of contacts moving southeast through the Philippine Sea. Reich, the boat's commanding officer, had been given intelligence that a Japanese naval task force was transiting the area — remnants of the fleet shattered weeks earlier at the Battle of Leyte Gulf. Through the squall and darkness, *Sealion*'s radar resolved the contacts into distinct returns — two large warships, consistent with capital ships, screened by several smaller escorts in a standard anti-submarine formation. Reich began his approach.

The larger of the two contacts was *Kongo* — 36,601 tons of Imperial Japanese Navy capital ship, a Kongo-class fast battleship commissioned in 1913 and modernized twice, armed with eight 14-inch guns in four twin turrets, capable of 30 knots, crewed by more than 1,200 sailors. *Kongo* represented a staggering investment of national resources. Her original construction at the Vickers shipyard in Barrow-in-Furness, England, had cost the Meiji-era Japanese government a significant portion of its naval budget. Her two subsequent modernizations — converting her from a battlecruiser to a fast battleship, strengthening her deck armor, upgrading her fire control — had consumed additional years of shipyard capacity and millions of yen. She was one of four Kongo-class battleships that had formed the backbone of Japanese surface power for three decades, and she had served in every major naval engagement of the Pacific War from the Indian Ocean raids to Leyte Gulf.

Sealion was a Balao-class fleet submarine, displacing 2,415 tons submerged — less than a fifteenth of *Kongo*'s displacement. She had been commissioned in October 1943 at the Electric Boat Company in Groton, Connecticut, at a cost of approximately $5.5 million. The Mark 18 electric torpedoes in her tubes cost roughly $10,000 each. At 0256 on November 21, Reich maneuvered *Sealion* into firing position and launched a spread of six torpedoes at the formation from a range of approximately 3,000 yards. Two struck *Kongo* amidships, blowing open her hull below the waterline. A third hit the destroyer *Urakaze*, which exploded and sank with all 228 hands.

Kongo listed heavily, and her crew fought the flooding for two hours. Damage control parties struggled to shore bulkheads and restore power. At 0524, her forward magazines detonated. She broke apart and sank in the Formosa Strait, taking approximately 1,200 of her crew with her. She was the only battleship sunk by an Allied submarine in the entire Pacific War — and a concrete demonstration that the most expensive warship in a formation could be destroyed by the cheapest weapon in the engagement.

The Young School and the Torpedo Revolution

The idea that cheap naval platforms could destroy expensive ones did not begin with submarines. It began with a theory, a French admiral, and a weapon that terrified every navy on Earth.

In the 1880s, Admiral Theophile Aube and a group of French naval theorists developed a doctrine they called the Jeune Ecole —

the Young School. Their argument was straightforward and radical. France could not outbuild the Royal Navy in battleships. Britain's industrial capacity, imperial wealth, and established shipyard infrastructure guaranteed superiority in capital ships for the foreseeable future. The conventional response was to accept permanent naval inferiority. Aube refused.

The recently invented self-propelled torpedo — first demonstrated by Robert Whitehead in 1866 and rapidly improved through the 1870s and 1880s — offered an alternative. Mated to small, fast, cheap torpedo boats, the torpedo could reach below the waterline of even the most heavily armored battleship. A torpedo boat costing a fraction of a battleship could, in theory, sink that battleship with a single hit. Build enough torpedo boats, Aube argued, and the battleship became a liability — an enormous concentration of national wealth that could be negated by a weapon costing one-hundredth as much.

The Jeune Ecole was, in modern terms, a cost-exchange theory. It recognized that the industrial economics of naval warfare had shifted with the torpedo's introduction. The torpedo did not need to match the battleship in firepower, armor, or crew — it needed only to put an explosive charge below the waterline. Theodore Ropp, whose history of French naval policy remains the definitive account of the period, documented how Aube's doctrine panicked the Royal Navy and launched a naval arms race that reshaped every fleet on Earth. The British Admiralty took the torpedo boat threat seriously enough to restructure its entire home defense strategy. Torpedo boats were cheap, fast, and — in the confined waters of the English Channel — potentially decisive against capital ships at anchor or in transit.

The British response was telling. They did not abandon the battleship. They did not restructure their fleet around the new economics. They built a new class of warship specifically designed to counter the torpedo boat: a fast, lightly armed vessel that could screen the battle fleet and hunt down torpedo boats before they reached firing range. They called it the torpedo boat destroyer. Within a decade, the name was shortened to "destroyer," and it had become a standard warship type in every major navy.

The Destroyer Trap

The destroyer is the critical link in the pattern. It represents the institutional response that recurs in every era of cost inversion: build an expensive counter-system to protect the expensive platform.

The torpedo boat cost almost nothing relative to a battleship. The destroyer that countered the torpedo boat cost far more than the torpedo boat — but less than the battleship it protected. On the surface, this appeared rational. The fleet now had a layered defense. But the total cost of naval power had increased. The fleet now required battleships *and* destroyers, plus the crews, the fuel, the maintenance, and the dockyards to sustain both. The cheap threat had not been defeated. It had been absorbed into a more expensive defensive architecture that preserved the battleship's centrality while adding the cost of its protection.

Martin Samuels, in his analysis of the Jeune Ecole's legacy, noted that the destroyer's success against torpedo boats was real but temporary. Within two decades, the torpedo boat had evolved into the submarine — a platform that was harder to detect, harder to destroy, and capable of carrying more torpedoes to greater range while operating from a concealed position beneath the surface. The destroyer, originally built to counter a surface threat, was repurposed to counter a submerged one. Anti-submarine warfare became the most expensive, most manpower-intensive, and most technologically demanding mission in naval operations — a distinction it would hold for the next century.

The cost of countering the cheap platform kept rising. The cost of the cheap platform kept falling relative to the value of what it could destroy. The destroyer had not solved the problem. It had institutionalized it.

The parallel to today is precise. The counter-drone systems now being fielded by the United States and its allies — directed-energy weapons, electronic warfare suites, kinetic interceptors, AI-enabled detection systems — are the destroyers of the drone age. They are built to protect expensive platforms from cheap threats. They cost more than the threats they counter. They will be effective against the current generation of cheap threats, just as the destroyer was effective against the torpedo boat. And they will face the same problem the destroyer faced. The cheap threat will evolve into something harder to counter, and the cost of the counter-system will continue to rise.

The Submarine's Ledger

Two world wars proved the Jeune Ecole more prescient than even Aube imagined — though the weapon was the submarine, not the torpedo boat he envisioned. In World War I, German U-boats demonstrated the strategic potential of cheap naval platforms against expensive ones with a brutality that shocked the world. On September 22, 1914 — barely seven weeks into the war — a single submarine, U-9, commanded by Kapitanleutnant Otto Weddigen, sank three British armored cruisers — HMS *Aboukir*, *Hogue*, and

Cressy — in less than an hour off the coast of the Netherlands, killing 1,459 sailors. The three cruisers displaced a combined 36,000 tons. U-9 displaced 611 tons. The cost ratio between the submarine and its victims was staggering.

Over the course of the war, German U-boats sank more than 5,000 Allied merchant ships and 10 major warships. The Allied response consumed vast resources — convoy systems that required hundreds of escort vessels, destroyer flotillas dedicated to anti-submarine patrol, depth charges developed and produced by the thousands, hydrophones and early sonar systems, dedicated patrol aircraft, and an entire anti-submarine doctrine built from scratch. The cost of countering the submarine exceeded the cost of the submarines themselves by orders of magnitude.

In World War II, the mathematics intensified. German U-boats sank 3,500 Allied merchant ships and 175 warships in the Battle of the Atlantic. American submarines sank 1,314 Japanese merchant vessels and 201 warships in the Pacific. The toll included aircraft carriers. USS *Archerfish* sank the Japanese carrier *Shinano* — at 71,890 tons, the largest warship ever sunk by a submarine — on November 29, 1944, just eight days after Reich's attack on *Kongo*. Theodore Roscoe's operational history of US submarine warfare documented that American submarines, representing less than 2% of US Navy personnel, accounted for more than 55% of all Japanese shipping destroyed in the war.

The cost-exchange ratio was devastating. A fleet submarine cost roughly $5-6 million. A battleship cost $70-100 million. An aircraft carrier cost $50-80 million. The torpedo that sank any of them cost $10,000. The mathematics of submarine warfare were, at their core, identical to the mathematics of Agincourt: a cheap weapon, deployed at scale by a relatively inexpensive platform, destroying targets that cost orders of magnitude more to build, crew, and sustain.

The response to the submarine threat in both wars consumed staggering resources. In World War II alone, the Allied anti-submarine warfare effort included thousands of escort vessels, hundreds of long-range patrol aircraft, dedicated anti-submarine carrier groups, code-breaking operations at Bletchley Park and elsewhere that employed thousands of mathematicians and linguists, sonar systems installed on every escort vessel, depth charges produced by the millions, and the convoy system itself — which slowed the delivery of vital cargo by requiring ships to sail in formation at the speed of the slowest vessel. The cost of defending against the submarine far exceeded the cost of the submarine fleet it was defending against, and the threat was never eliminated — only managed, at escalating cost, until the war ended.

The V-1 and V-2: When the Expensive Side Gets the Economics Backward

The naval pattern was not the only cost-exchange lesson of the Second World War. Germany's vengeance weapons program produced a case study so precise in its economic implications that defense economists still cite it eight decades later.

The V-1 flying bomb was a crude, cheap, mass-producible weapon. It used a simple pulse-jet engine that consumed low-grade fuel, carried an 1,870-pound warhead, and could be manufactured by semi-skilled labor in dispersed factories. Its cost was approximately 3,500 Reichsmarks per unit — roughly $35,000 in 2025 dollars. It was not accurate. It was not fast. Allied fighters could intercept it, and anti-aircraft gunners could shoot it down. Between June 1944 and March 1945, Germany launched more than 9,500 V-1s against England and approximately 12,000 against targets on the European continent. The V-1 was, in essence, a mass-production weapon — designed for quantity, not quality.

The V-2 ballistic missile was the opposite in every respect: a technological marvel and an economic catastrophe. It traveled faster than the speed of sound, reaching altitudes above 50 miles before descending on its target. No contemporary defense could intercept it. It carried a slightly larger warhead — 2,150 pounds — but cost approximately 100,000 Reichsmarks per unit, nearly 30 times the V-1's price. It required specialized materials, precision-machined components, advanced guidance systems, liquid oxygen oxidizer that had to be produced and transported under exacting conditions, and a vast underground production facility at Mittelwerk that consumed enormous quantities of slave labor and critical engineering talent.

Adam Tooze, in *The Wages of Destruction*, calculated that the V-2 program consumed resources equivalent to building several panzer divisions — armored formations that could have materially affected the ground war on the Eastern Front. The total investment in V-2 development and production, including the Peenemunde research facility and the Mittelwerk factory, represented one of the largest single weapons programs in the Third Reich, consuming steel, aluminum, skilled labor, and industrial capacity that Germany desperately needed elsewhere.

The comparison is stark. The V-1 delivered more total explosive tonnage to England than the V-2, at a fraction of the cost per warhead delivered. When the metric is explosive tonnage on target per Reichsmark invested, the cheap weapon won by a wide margin. Germany invested massively in the expensive, technologically sophisticated weapon at the direct expense of the cheap, mass-producible one — driven by institutional prestige, engineering ambition, and the political appeal of a "wonder weapon." Tooze's economic analysis demonstrated that this choice was among the most consequential resource misallocations of the war.

The V-1 and V-2 are not a story about naval warfare. They are a story about what happens when an institution chooses the expensive weapon over the cheap one because the expensive weapon is more prestigious, more technically impressive, and more aligned with the institution's self-image. The parallel to a military that prefers a $170 million stealth fighter over a $500 drone does not require elaboration.

The Same Pattern, Every Time

Three cases across two domains, and the conclusion is the same.

At Agincourt, the cheap weapon was the longbow and the arrow. At sea, it was the torpedo and the submarine. In the air, it was the pulse-jet flying bomb. In each case, the expensive side faced the same choice — adapt to the new economics or build a counter-system that preserved the old ones. In each case, the institution chose the counter-system. The French built heavier armor after Crecy; it did not save them at Agincourt. The Royal Navy built destroyers; the submarine rendered them insufficient. The Allied air forces built elaborate anti-V-1 defenses — barrage balloons, anti-aircraft batteries, fighter patrols — that consumed far more resources than the V-1s they shot down.

The pattern holds because the economics hold. When a cheap weapon can reliably threaten an expensive platform, the defender faces a structural disadvantage that no single counter-system can resolve. The defender must protect the platform every time. The attacker needs to succeed only once. The defender's counter-system costs more than the threat it counters. The attacker iterates — new tactics, new delivery methods, new variants — at a fraction of the defender's adaptation cost. The cost curves diverge. They always diverge.

The battleship was the aircraft carrier of the 1940s. The submarine was the Houthi drone. The expensive side built a counter-system and called it progress. The cheap side kept building torpedoes.

The torpedo boat became the submarine. The anti-submarine warfare mission became the most expensive single category of naval expenditure. Each step in the escalation cost more on the defending side than the attacking side. The most recent version of this pattern did not happen at sea. It happened on the roads of Iraq and Afghanistan — and it cost the United States $45 billion.

Chapter 7: The $45 Billion Road

Corporal Todd Nicely lost both legs and both arms on a dirt road in Helmand Province, Afghanistan, on March 26, 2010.

He was a rifleman with 2nd Battalion, 2nd Marine Regiment, on a foot patrol outside Forward Operating Base Geronimo in the Marjah district, moving through a landscape of packed earth, irrigation ditches, and poppy fields where every step was a deliberate calculation. The improvised explosive device that detonated beneath his feet was built from a yellow plastic jug of the kind sold at any bazaar in southern Afghanistan, packed with ammonium nitrate fertilizer — the same compound used to grow the poppy crop surrounding the patrol route — and wired to a pressure-plate detonator fashioned from two strips of wood, a hacksaw blade, and a clothespin. The total cost of the device, according to estimates compiled by the Joint Improvised-Threat Defeat Organization, was between $20 and $100.

Nicely survived. He became one of only two living quadruple amputees in the United States Marine Corps. His medical care, rehabilitation, and prosthetics would eventually cost the Department of Veterans Affairs millions of dollars — a sum that did not appear in any accounting of the counter-IED program, but that was as much a product of that $20 weapon as the crater it left in the road.

The mine-resistant ambush-protected vehicle — the MRAP — that other Marines in his unit rode in cost between $500,000 and $1 million, depending on the variant. It was a purpose-built armored truck, engineered with a V-shaped hull to deflect blast energy away from the crew compartment, fitted with ballistic glass rated to stop .50-caliber rounds, reinforced with composite armor panels, and mounted on a mine-resistant suspension system. The vehicle represented tens of thousands of engineering hours, a dedicated production line at facilities in Wisconsin, South Carolina, and Pennsylvania, and a logistics chain stretching from factories in the American heartland to airfields in Kuwait and forward operating bases across southern Afghanistan.

The MRAP was the most visible product of the largest counter-weapons program in American military history. By 2010, the United States had spent more than $45 billion on counter-IED efforts — a figure that encompassed not only the 27,000 MRAPs purchased at a total program cost exceeding $50 billion, but the electronic countermeasures mounted on every patrol vehicle, the route-clearance teams that swept roads before convoys, the intelligence fusion cells that tracked bomb-making networks, the forensic exploitation laboratories that analyzed recovered IED components for fingerprints and DNA, and the institutional architecture of an entire organization — the Joint IED Defeat Organization, or JIEDDO — created solely to counter a weapon that cost less than a dinner for two at a Kabul restaurant.

The Weapon That Won

The improvised explosive device was the most tactically significant weapon of the first two decades of the twenty-first century. This is not a subjective assessment. It is a statement of statistical fact.

In Afghanistan, IEDs caused more than half of all coalition casualties — by some estimates 50 to 60 percent of soldiers killed or wounded by hostile action fell to buried bombs, not bullets. In

Iraq, the figure was comparable. Between 2001 and 2021, IEDs killed or wounded more coalition soldiers than every other weapon system combined — more than rifles, more than rocket-propelled grenades, more than mortars, more than vehicle-borne ambushes, more than sniper fire. No other single weapon type in any conflict since the Korean War accounted for such a dominant share of casualties among a major military power's deployed forces.

The Government Accountability Office, in a series of reports spanning the wars in Iraq and Afghanistan, documented the escalating lethality of IEDs and the corresponding escalation in US counter-IED spending. The numbers trace a cost curve that defense economists recognized immediately as unsustainable. In fiscal year 2004, JIEDDO's budget was approximately $1.3 billion. By fiscal year 2006, it had nearly doubled. By fiscal year 2010, annual spending reached $3.5 billion. By the time the organization was redesignated as the Joint Improvised-Threat Defeat Organization (JIDO) in 2016, cumulative spending on counter-IED programs had exceeded $45 billion. Some estimates, including the full cost of the MRAP program and associated logistics, training, maintenance, and medical care for IED casualties, place the total closer to $75 billion.

The weapon this money was designed to defeat cost between $20 and $100 to build. A jug. Fertilizer. Wire. A battery.

The Cost-Exchange Arithmetic

The cost-exchange ratio between an IED and its countermeasures is the starkest in modern military history before the drone — and the most direct precedent for the cost inversion unfolding today. Consider the arithmetic at the individual engagement level. A $30 IED can destroy or disable a $700,000

MRAP. If the MRAP survives — as it was designed to — the vehicle still requires inspection, damage assessment, and often replacement of blown tires, damaged axles, cracked ballistic glass, and scorched armor panels costing tens of thousands of dollars per incident. The route must be re-cleared by a route-clearance team operating specialized vehicles that cost millions. The patrol must be rerouted. Intelligence resources must be redirected to identify the emplacement network. Medical evacuation assets must be positioned. The engagement cost to the defender, even in a "successful" defense where no one dies, runs to hundreds of thousands of dollars per incident. The engagement cost to the attacker was $30.

Now consider the system-level arithmetic. The United States spent $45 billion or more on counter-IED programs. The Taliban and Iraqi insurgents spent a tiny fraction of that amount on the IEDs themselves. Andrew Krepinevich and Barry Watts at the Center for Strategic and Budgetary Assessments estimated that the cost-exchange ratio across the full counter-IED enterprise ran between 100:1 and 1,000:1 — the United States was spending one hundred to one thousand dollars in countermeasures for every dollar the adversary spent on the weapon.

These ratios should sound familiar. They are the same ratios now emerging in the counter-drone domain. An approximately $4.9 million SM-6 interceptor missile fired at a $35,000 Houthi drone produces a cost-exchange ratio of approximately 140:1. A $1.2 million Evolved SeaSparrow Missile fired at a $2,000 FPV drone produces a ratio of 600:1. The IED established the template. The drone is the successor.

The Adaptation Cycle

What made the IED problem intractable was not the weapon itself — a buried bomb is conceptually one of the oldest weapons in warfare. What made it intractable was the adaptation cycle — the speed at which the cheap side could iterate relative to the expensive side.

The United States deployed electronic countermeasures — vehicle-mounted jammers that broadcast across radio frequencies — to disrupt remote-controlled IED detonators. Each vehicle jammer cost tens of thousands of dollars, plus installation, maintenance, and the power draw on the vehicle's electrical system. The insurgents switched to command-wire detonation, running a physical wire from the device to a triggering position hundreds of meters away, rendering the radio-frequency jamming equipment useless. The adaptation cost a spool of wire — less than $5.

The US deployed more sophisticated jammers covering wider frequency bands, including systems that could detect and neutralize cell-phone-triggered devices, at a cost of hundreds of millions of dollars across the fleet. The insurgents switched to pressure-plate detonators — purely mechanical triggers activated by the weight of a vehicle or a footstep, requiring no electronic signal at all. No frequency to jam. No signal to detect. No electronic signature to trace. Each generation of pressure plate cost less than $5 to build from materials available at any hardware market in Kandahar.

The US deployed MRAPs with V-shaped hulls to survive underbody blasts. The insurgents increased charge sizes, using multiple jugs of ammonium nitrate stacked together. They shifted to explosively formed penetrators — EFPs — shaped copper liners backed by explosive that, upon detonation, collapsed the copper into a superplastic slug traveling at two kilometers per second, capable of punching through the side armor of any vehicle in the coalition inventory. The copper liner cost dollars. The explosive cost dollars. The slug it formed could defeat armor that cost hundreds of thousands of dollars to install.

The US deployed route-clearance teams with ground-penetrating radar, mine-detection dogs trained to detect ammonium nitrate, and specialized engineering vehicles costing millions each. The insurgents buried devices deeper, used non-metallic containers that defeated metal detectors, sealed charges in plastic wrap to reduce the scent signature that defeated the dogs, and emplaced multiple devices in patterns designed to target the clearance teams themselves — secondary devices positioned to strike the vehicles that stopped to investigate the primary.

Each US adaptation cost millions to develop, test, field, train on, and sustain in theater. Each insurgent adaptation cost dollars. Each cycle widened the gap.

The structural logic was inescapable. The side spending more was constrained by procurement cycles, testing requirements, safety certifications, training programs, and logistics chains that stretched from the Pentagon to defense contractors to testing ranges to theater logistics commands. A new electronic countermeasure required an engineering proposal, a contract award, a development phase, testing at proving grounds in the United States, integration onto vehicles, training for operators and maintainers, and shipping to theater — a process that consumed 12 to 18 months in the best case. A new pressure-plate design required a trip to the hardware market and an afternoon's labor.

The military historian Thomas Rid, analyzing technology and irregular warfare, documented what he termed the adaptation paradox: the more sophisticated the defending force, the slower it adapts, because sophistication requires process and process requires time. The insurgent adapts by changing a component. The counterinsurgent adapts by changing a system. The system change is always slower and always more expensive. The gap between the two adaptation speeds compounds with every cycle, and each cycle leaves the expensive side further behind.

The Institutional Response

JIEDDO — the Joint IED Defeat Organization — was the Pentagon's answer to the IED crisis, and its history illustrates the institutional dynamics that recur in every cost inversion.

Deputy Secretary of Defense Gordon England created JIEDDO in 2006 as a director-level organization reporting directly to his office, and it grew rapidly into one of the largest single-mission organizations in the Department of Defense. At its peak, it employed more than 1,800 people — military, civilian, and contractor — and managed an annual budget exceeding $3.5 billion. It funded hundreds of technology programs, from electronic countermeasures to biometric identification systems that could match a thumbprint on a recovered IED component to a detainee in a holding facility, to predictive analytics tools that attempted to identify IED emplacement networks before the devices were built, to specialized forensic laboratories that subjected recovered bomb fragments to the same analytical rigor that the FBI applied to domestic crime scenes.

Some of these programs worked. The MRAP saved lives — an assessment supported by multiple studies, including a RAND Corporation analysis, showing significantly lower killed-in-action rates for troops riding in MRAPs compared to the up-armored

Humvees they replaced. Electronic countermeasures forced the insurgents to adopt less reliable detonation methods, reducing the overall success rate of radio-controlled devices. Intelligence-driven targeting disrupted bomb-making cells in both Iraq and Afghanistan, capturing or killing key facilitators in the IED supply chain. JIEDDO was not a failure in operational terms.

It was a failure in economic terms. The organization spent $45 billion to suppress — not defeat — a weapon that cost almost nothing. The IED was never eliminated. It was managed, at enormous and escalating cost, for as long as coalition forces remained in the theater. When forces withdrew, the IED remained. The Taliban continued to use it against Afghan government forces. Other groups in other theaters adopted it. The weapon's simplicity, low cost, and effectiveness ensured its survival long after the counter-program that fought it was dissolved.

The counter-IED enterprise could not reverse the fundamental cost equation. The defender's cost curve rose faster than the attacker's. Every new countermeasure generated a cheaper counter-countermeasure. The adaptation race was structurally asymmetric, and the asymmetry favored the attacker.

Andrea Dew and Marc Genest, writing in the *Journal of Strategic Studies*, described JIEDDO as "the most expensive single-threat counter-program in military history" and argued that its trajectory revealed a structural truth about asymmetric warfare: the defender's cost curve always rises faster than the attacker's, because the defender must protect everything while the attacker needs only to find one vulnerability. The attacker chooses the time, the place, and the method. The defender must be ready everywhere, all the time, against every variant. This asymmetry is not a feature of the IED specifically. It is a feature of any conflict in which a cheap weapon threatens an expensive target.

When the Pentagon redesignated JIEDDO as JIDO in 2016, the organization's own after-action assessments acknowledged what the economics had shown for years: the counter-IED mission would persist as long as the conditions that made IEDs attractive persisted. Those conditions — cheap materials, simple construction, devastating effect against expensive targets, and an adaptation cycle that favored the builder over the defender — were not going away. They were, if anything, intensifying as the technology of cheap weaponry continued to advance.

The Template for Today

The IED cost twenty dollars. The counter-IED program cost forty-five billion. The IED won.

This is not a judgment about the dedication or competence of the soldiers, Marines, engineers, intelligence analysts, and scientists who fought the IED problem. Many of them saved lives through extraordinary effort and ingenuity. The judgment is about the economics. A $20 weapon that forces a superpower to spend $45 billion in countermeasures — and that persists despite those countermeasures, continuing to inflict casualties until the last day of the last deployment — has achieved strategic effect vastly disproportionate to its cost. The attacker's total investment across the entire campaign was measured in tens of thousands of dollars. The defender's investment was measured in tens of billions.

The drone cost inversion now unfolding in Ukraine, the Red Sea, and beyond follows the identical structural logic. A $200 FPV drone forces the deployment of counter-UAS systems costing millions. A $35,000 Houthi drone forces the expenditure of an approximately $4.9 million interceptor. The adaptation cycle has already begun: fiber-optic tethered drones that defeat radio-frequency jamming, visual-navigation systems that operate without GPS in denied environments, coordinated multi-drone attacks that saturate point-defense systems designed for single-target engagements. Each new drone variant costs hundreds of dollars to produce. Each new counter-system costs millions to develop and field.

The IED was the rehearsal. The drone is the performance. And the cost-exchange math is worse — because drones can strike from any direction, at any altitude, against any target type, with a precision that the IED, confined to roads and footpaths, could never achieve. The IED was a passive weapon — it sat and waited for its target to come to it. The drone is an active weapon — it seeks its target wherever the target hides. The IED could threaten convoys and foot patrols. The drone can threaten tanks, ships, aircraft on runways, radar installations, command posts, logistics depots, and anything else that costs more than the drone itself. The geography of vulnerability has expanded from a road in Helmand to everywhere.

Agincourt. The torpedo. The IED. Each one inverted the cost equation and forced the expensive side into a losing race against a cheap adversary that could adapt faster and iterate cheaper. But one historical parallel cuts closer to today's drone math than any other — because it involves the precise mechanism of defense cost outrunning offense cost so completely that the entire strategy became economically unsustainable. That parallel sits in the Pentagon's own files, in a cost-exchange model that ended not with a battle, but with a treaty.

Chapter 8: The Pattern

Alain Enthoven was thirty-seven years old and running the numbers that would end a weapons program.

It was early 1968, and Enthoven, the Assistant Secretary of Defense for Systems Analysis under Robert McNamara, sat in his office in the E Ring of the Pentagon reviewing cost-exchange models for the Nike-X anti-ballistic missile system. The models had been built by his analysts at the Office of Systems Analysis — the group McNamara had created to impose economic rigor on defense decisions that had previously been driven by institutional preference and interservice politics. Enthoven was a RAND Corporation alumnus, a disciple of Charles Hitch, and a believer in the proposition that defense planning was, at its core, an economics problem. The question his team had been asked to answer was straightforward. Could the United States defend its population against a Soviet nuclear strike at a cost it could sustain?

The answer was no. And the math was precise enough to prove it.

The Nike-X system proposed to ring American cities with batteries of Sprint and Spartan interceptor missiles, backed by phased-array radars that could track hundreds of incoming warheads simultaneously. The technology was ambitious. The engineering was plausible. The cost, according to initial estimates, was approximately $30-40 billion in 1968 dollars — a staggering sum, but not obviously prohibitive for a nation spending $80 billion a year on defense during Vietnam. The question Enthoven's team asked was not whether the system could work. It was whether the system could work *economically* — whether the cost of defending against a Soviet attack could be sustained against a Soviet response.

Enthoven's analysts modeled the cost-exchange ratio between offense and defense — what it would cost the United States to intercept an additional Soviet warhead versus what it would cost the Soviet Union to add another warhead to overwhelm the defense. The results were consistent across every scenario his team ran. The United States would need to spend between $2 and $20 in defensive interceptors and associated infrastructure for every $1 the Soviets spent on offensive missiles. At a target of limiting casualties to 10% of the population in a full exchange — the most optimistic credible assumption — the ratio stabilized at approximately 6:1. When corrected for the purchasing-power differential between dollars and rubles — Soviet missiles cost less in real resources than American interceptors because Soviet labor, materials, and manufacturing were cheaper — the ratio jumped to 20:1 or higher.

The implication was inescapable. No matter how effective the interceptors, no matter how advanced the radar, no matter how fast the computers that directed the engagement, the Soviets could always add warheads for less money than the Americans could add interceptors. Build a defense, and the offense adds warheads. Expand the defense, and the offense adds more warheads. At every step, the offense spends less than the defense. The cost curves diverge, and they diverge in the attacker's favor. Strategic missile defense was economically unsustainable — not because the technology failed, but because the economics guaranteed that the attacker would always win the spending race.

The Treaty That Proved the Math

Enthoven and his co-author K. Wayne Smith documented the analysis in *How Much Is Enough?*, published in 1971. The book became the foundational text of defense cost-exchange analysis — the methodology of comparing the marginal cost of an additional unit of offense against the marginal cost of an additional unit of defense. Four years after Enthoven's analysis reached the Secretary of Defense's desk, the ABM Treaty of 1972 — one of the most consequential arms control agreements of the Cold War — codified the conclusion.

The treaty limited both the United States and the Soviet Union to a single ABM deployment site each — later reduced from the original two sites per side by the 1974 Protocol. The United States chose to defend a single ICBM field at Grand Forks, North Dakota. The Soviets chose to defend Moscow. Neither side attempted to defend its population, because Enthoven's math — confirmed independently by Soviet analysts running their own models — demonstrated that population defense was economically impossible against a determined adversary willing to spend on offense.

The ABM Treaty was not a triumph of diplomacy in the usual sense. It was a recognition of arithmetic. Both superpowers looked at the cost-exchange ratio, recognized that the defense could never win the spending race against the offense, and agreed to stop trying. The most powerful nations on Earth, armed with thousands of nuclear warheads, accepted mutual vulnerability because the alternative — trying to buy invulnerability — was mathematically guaranteed to fail.

The Universal Equation

The ABM cost-exchange analysis was the first time a major military decision was explicitly, quantitatively driven by the kind of cost-ratio arithmetic that this book has traced across six centuries. But the pattern Enthoven quantified did not begin in 1968. It began at Agincourt.

Strip the four preceding cases to their structural elements, and the same equation emerges every time. At Agincourt, cheap longbowmen destroyed expensive mounted knights at a cost-exchange ratio so extreme that the English investment per French casualty was negligible. The longbow cost shillings. The knight cost a fortune. The arrows were mass-produced by fletchers working to standardized specifications. The armor was bespoke, hand-fitted by master craftsmen in Milan and Augsburg. The battle lasted hours. The aristocratic military class that defined itself through mounted combat took decades to accept what the field at Azincourt had demonstrated.

At sea, cheap torpedo boats — and their successors, submarines — sank capital ships representing astronomical concentrations of national wealth. The torpedo cost $10,000. The battleship cost $70-100 million. The response was the destroyer, then the convoy system, then anti-submarine warfare doctrine — each layer of defense costing more than the last, none of them eliminating the threat. Germany's V-1, a crude pulse-jet flying bomb costing 3,500 Reichsmarks, delivered more explosive tonnage per unit of investment than the V-2 ballistic missile costing 100,000 Reichsmarks — a demonstration that the expensive, prestigious weapon was not the economically rational one.

On the roads of Iraq and Afghanistan, a $20 IED forced a $45 billion counter-program. The cost-exchange ratio across the full enterprise ran between 100:1 and 1,000:1. Every American adaptation — electronic countermeasures, MRAPs, route clearance, intelligence fusion — was met by a cheaper insurgent counter-adaptation. Pressure plates that defeated jammers. Stacked charges that defeated armor. Secondary devices that defeated clearance teams. The weapon was never defeated. It was managed, at enormous and escalating cost, until the forces withdrew. And in 1968, Enthoven's models proved that the same equation applied at the strategic level: the offense always won the cost race against the defense, at ratios between 2:1 and 20:1.

The Five Elements

The pattern is not merely recurring. It is structural. Every case shares five elements, and no case across six centuries deviates from the sequence.

First, a cheap weapon appears that can reliably destroy or neutralize an expensive one. The longbow killed knights. The torpedo sank battleships. The IED destroyed armored vehicles. The ICBM overwhelmed missile defenses. In each case, the cheap weapon did not need to be technologically superior to its target. It did not need to be more accurate, more durable, or more versatile. It needed only to deliver a lethal or neutralizing effect at a fraction of the target's cost.

Second, the expensive side builds a counter-system. The French built heavier armor after Crecy. The Royal Navy built torpedo boat destroyers. The US military built MRAPs and electronic countermeasures. The Pentagon proposed Nike-X interceptors. The counter-system is always framed as the restoration of the old order — the technological fix that will neutralize the cheap threat and preserve the expensive platform's dominance. The counter-system is always funded generously, because the institution that operates the expensive platform controls the budget.

Third, the counter-system costs more than the threat it counters. The destroyer cost more than the torpedo boat. The MRAP cost more than the IED. The interceptor cost more than the warhead. This is not an accident. It is structural. A counter-system must detect, track, and neutralize a threat — three functions that are inherently more complex and more expensive than the single function of the threat itself, which is to deliver a destructive effect to a target. Defense is always more expensive than offense when the offense is cheap, because the defender must solve a harder problem.

Fourth, the cheap side iterates faster because iteration is cheaper. The Iraqi insurgents switched IED trigger mechanisms in days. The Soviets could add warheads faster and cheaper than the Americans could add interceptors. English fletchers produced arrows in weeks; Milanese armorers needed months to produce a single suit of plate. Each cycle of adaptation compounds the cost disadvantage of the expensive side, because the expensive side's adaptation cycle is constrained by procurement processes, testing requirements, safety certifications, training programs, and logistics chains, while the cheap side is constrained by almost nothing except the availability of basic materials and the ingenuity of its engineers or fighters.

Fifth — and this is the element that transforms a tactical problem into a strategic one — the institution that built its identity around the expensive platform resists adaptation longest. The French aristocracy could not abandon the mounted knight because knighthood was their identity, their social position, their reason for existing as a class. The navies of the world could not abandon the battleship because admirals had built their careers commanding them, shipyards had built their businesses constructing them, and the entire culture of naval power was organized around their operation. The Pentagon could not abandon the MRAP and its associated counter-IED enterprise because JIEDDO had become an institutional constituency with 1,800 employees, a $3.5 billion annual budget, and a director who reported to the Deputy Secretary of Defense.

In every case, the last obstacle to adaptation was not technical. It was organizational.

The Cost-Ratio Table

Place the cases side by side, and the escalation across six centuries becomes visible — and the current moment comes into focus. At Agincourt, the cost ratio between the archer's arrow and the knight's armor was effectively infinite — the arrow cost almost nothing, the knight cost a fortune. Express it conservatively as a ratio of total force costs, and the English investment per French casualty was a small fraction of the French investment per English casualty. Call it 50:1 as a rough order of magnitude, acknowledging that precise medieval cost accounting is impossible.

The torpedo-to-battleship ratio was narrower but still extreme. A $10,000 torpedo sinking a $70-100 million battleship produces a ratio between 7,000:1 and 10,000:1. Even accounting for the cost of the submarine that delivered the torpedo — $5-6 million — the ratio exceeds 10:1 for the delivery platform alone.

The IED-to-MRAP ratio at the individual engagement level ran $30 versus $700,000, or approximately 23,000:1. At the program level — $45 billion in counter-IED spending versus perhaps $50 million in total insurgent IED construction costs across the entire Afghan campaign — the ratio approaches 1,000:1 in favor of the attacker.

The ABM ratio, as Enthoven calculated it: 2:1 to 20:1, favoring the offense at the strategic level.

Now consider the drone. A $200 FPV drone destroying a $4.2 million T-90M main battle tank produces a cost-exchange ratio of 21,000:1. An approximately $4.9 million SM-6 interceptor fired at a $35,000 Houthi drone produces a defensive cost-exchange ratio of 140:1 — meaning the defender spends 111 times what the attacker spent on the threat. A $1.2 million Evolved SeaSparrow Missile fired at a $2,000 commercial-component FPV drone produces a ratio of 600:1. These are not projections. They are observed ratios from ongoing military operations in Ukraine and the Red Sea.

The current cost inversion is the most extreme in recorded military history. No previous case — not the longbow, not the torpedo, not the IED, not the ICBM — produced offensive cost-exchange ratios as favorable to the cheap side as the FPV drone produces today. The 21,000:1 ratio between a $200 drone and a $4.2 million tank is not merely worse than the ABM problem that forced a treaty between superpowers. It is worse by three orders of magnitude.

Why Institutions Fail the Test

The five-element pattern explains the mechanics of cost inversion. It does not, by itself, explain why the expensive side keeps making the same mistake across centuries, across domains, across entirely different institutional contexts. For that, the explanation lies in institutional economics — the study of how organizations make decisions that serve the organization's survival even when those decisions undermine the organization's stated mission.

The defense economist Franklin Spinney, whose work on defense budget reform made him the most prominent internal critic the Pentagon produced in the twentieth century, identified this dynamic as the "plans/reality mismatch" — the institutional tendency to plan around the platforms the organization wants to build rather than the threats the organization actually faces. Spinney argued, in testimony before Congress and in his landmark study *Defense Facts of Life*, that defense budgets are not driven primarily by threat analysis. They are driven by the institutional need to sustain production lines, protect congressional-district jobs, maintain career tracks for officers whose promotions depend on managing specific weapons programs, and preserve the organizational structures that the current generation of leaders built its careers within. The platform comes first. The threat assessment is adjusted to justify the platform.

James Q. Wilson, in *Bureaucracy*, described the same dynamic from the perspective of organizational theory. Organizations define themselves by their "core tasks" — the activities around which they organize their personnel, their budgets, their training, and their promotion systems. The Army's core task is combined-arms maneuver. The Navy's core task is power projection from carrier strike groups. The Air Force's core task is air superiority through crewed fighter aircraft. Suggest that these core tasks are becoming less relevant — that cheap drones might replace expensive tanks, that swarms might threaten carriers, that autonomous aircraft might outperform piloted ones — and you are not making a technical argument. You are threatening the organization's identity. Organizations resist identity threats more fiercely than they resist external enemies.

This is why the pattern repeats with such consistency across centuries that have nothing else in common. The institution does not fail to see the threat. French commanders at Agincourt knew the longbow's lethality; they had seen it at Crecy and Poitiers. Pentagon leaders today know the drone's cost advantage; they have watched the footage from Ukraine. The institution fails to act on the threat, because acting on the threat requires dismantling the institutional structures built around the previous era's dominant platform. The French aristocracy could not dismantle the feudal military system that produced the knight. The Pentagon cannot dismantle the defense-industrial base that produces the F-35. Each institution sees the cost-exchange math clearly. Each institution chooses to manage the problem within existing structures rather than restructure around the new economics.

The Drone Parallel, Made Explicit

The mounted knight. The battleship. The MRAP. The aircraft carrier. Same platform, different century. Same mistake — spending more money on a weapon that is getting cheaper to kill.

The United States military is currently organized around expensive platforms — the F-35 at a program unit cost of $170 million per aircraft, the Ford-class carrier at $13 billion, the Columbia-class submarine at $9 billion, the Abrams tank at $10 million — that represent the pinnacle of an eight-decade military-industrial paradigm. These platforms are not failing in the traditional sense. They are the most capable weapons systems ever fielded by any military in history. The F-35's sensor fusion is unmatched. The Ford-class carrier's electromagnetic launch system is a genuine engineering achievement. The Columbia-class will carry the most survivable leg of the nuclear triad for decades.

They are also, by the pattern established across six centuries of evidence, the wrong investment for the emerging threat environment. When a $200 drone can destroy a $4.2 million tank, the economics of armored warfare have inverted. When a swarm of $35,000 drones can threaten a $13 billion carrier, the economics of naval power projection have inverted. When a $500 FPV drone can, theoretically, ground a $170 million fighter by striking its engine intake on the runway before it ever takes off, the economics of air superiority have inverted.

The Pentagon's response to date follows the pattern with precision that would be impressive if it were not so predictable. The counter-drone budget — $3.187 billion in the National Defense Authorization Act for fiscal year 2026, plus a $7.5 billion line in the Department of Defense budget — is the destroyer of the drone age. It is an expensive counter-system designed to protect expensive platforms from cheap threats. It may work tactically, as the destroyer worked against the torpedo boat and the MRAP worked against the IED. It will not reverse the cost curve, because the cost curve has never been reversed by a counter-system in six hundred years of military history. The cheap side will iterate — fiber-optic drones that defeat radio-frequency jamming, AI-guided navigation that operates without GPS, swarm tactics that saturate point defenses — and the cost of defense will continue to rise faster than the cost of offense.

The ABM Treaty recognized this reality for nuclear weapons in 1972. No equivalent recognition exists for the drone cost inversion — and the ratios today are three orders of magnitude worse than the ratios that forced two superpowers to accept mutual vulnerability.

What the Pattern Demands

The pattern does not demand fatalism. It demands honesty about the economics — the same honesty that Enthoven brought to the Nike-X debate in 1968.

Every military revolution is a cost revolution. The longbow was not a revolution in lethality — arrows had killed men for millennia before Agincourt. It was a revolution in the cost of lethality: the cost of killing a knight dropped from the price of another knight to the price of a handful of arrows. The torpedo was not a revolution in naval combat — ships had been sunk since Salamis. It was a revolution in the cost of sinking a ship: the cost dropped from another ship of equal size to a weapon that cost a thousandth as much. The IED was not a revolution in roadside ambush — guerrillas had attacked convoys for centuries. It was a revolution in the cost of the ambush relative to the cost of the target.

The drone is a revolution in the cost of precision strike. A $200 platform guided by a $50 camera and a $20 radio link can deliver a shaped-charge warhead to a specific vehicle, a specific bunker aperture, a specific radar antenna — with a precision that, a generation ago, required a cruise missile costing $1 million or more. The cost of precision has collapsed by four orders of magnitude. The institutions built around expensive precision — the aircraft carriers that launch precision strikes, the stealth fighters that deliver precision-guided munitions, the guided-missile destroyers that fire precision cruise missiles — face the same structural challenge that the mounted knight faced at Agincourt. Their defining capability is now available to anyone with a few hundred dollars and a soldering iron.

The pattern is clear. The history is unambiguous. But history is not destiny — it is a warning. The warning is not that expensive platforms are useless. The warning is that institutions which organize themselves around expensive platforms, and which respond to cheap threats by building expensive counter-systems, will lose the cost race every time. The only strategy that has ever worked against a cost inversion is to embrace the cheap side's economics — to build cheap, to build many, to accept losses, and to iterate faster than the adversary can adapt.

The question is whether the institution that spends $892.6 billion a year on defense is capable of reading the warning signs that six centuries of military history have posted in plain view. The answer is in the Pentagon's budget — in what it buys, what it fails to buy, and who benefits from the growing mismatch between the threat the nation faces and the money it spends. The numbers are not encouraging.

Chapter 9: The $892 Billion Question

Senator Roger Wicker held up two photographs.

The hearing room in the Hart Senate Office Building was full — standing room only along the back wall, staffers pressed shoulder to shoulder behind their principals, a half-dozen cameras from C-SPAN and defense media outlets angled toward the witness table. It was March 2026, and the Senate Armed Services Committee was conducting its annual posture hearing on the Department of Defense budget request. Wicker, the committee's chairman and a Mississippi Republican who had spent two decades on the panel, had seen dozens of these hearings. He had watched secretaries of defense from both parties present budgets measured in hundreds of billions, each one larger than the last, each one defended with the same language about modernization, readiness, and great-power competition. He had voted for most of them.

But this time, Wicker did something his colleagues did not expect. He set aside his prepared remarks and picked up two printed photographs from the desk in front of him. He held the first toward the witness table, where Deputy Secretary of Defense Stephen Feinberg sat flanked by the service chiefs. The photograph showed an F-35A Lightning II, the single most expensive weapons program in the history of the Department of Defense, parked on a flight line at Hill Air Force Base in Utah. The cost per aircraft was approximately $80 million for the latest Lot 18 production variant, with a lifetime operating cost that the Government Accountability Office estimated at more than $400 million per airframe across the F-35 program's projected service life.

Then Wicker held up the second photograph. It showed a first-person-view drone of the type that Ukrainian forces had been using to destroy Russian armored vehicles, artillery pieces, and supply trucks for nearly three years. The particular model in the photograph — a racing-style quadcopter with a 3D-printed tail assembly and an RPG warhead zip-tied to its undercarriage — cost approximately $500 to build from commercially available components.

"How many of these," Wicker said, holding the drone photograph higher, "does it take to ground one of these?" He raised the F-35 photograph in his other hand.

The room went quiet. The math was not complicated. At $500 per drone and $80 million per F-35, the ratio was 160,000 to one. Even if only one in a hundred drones reached its target, the attacker could afford to send 1,600 drones for the cost of a single aircraft. The question had no comfortable answer, because the comfortable answer — that the F-35 operates in a different threat envelope, that air-to-air combat and close air support cannot be reduced to a cost ratio, that the comparison is unfair — was true and irrelevant at the same time. The drone was not trying to fight the F-35 in the air. It was trying to destroy it on the ground, in the maintenance hangar, on the runway, at the fuel depot. And at those price points, the math did not care about threat envelopes.

Nobody at the witness table volunteered an answer. That silence contained the most dangerous economic truth in the United States defense budget.

Twenty Dollars for the Last War

The fiscal year 2026 defense budget request totaled $892.6 billion. It was the largest peacetime defense budget in American history, adjusted for inflation. It was also, by the Pentagon's own internal metrics, insufficient — the service chiefs testified that unfunded requirements exceeded $30 billion, and the Congressional Budget Office projected that maintaining the current force structure through 2034 would cost an additional $157 billion above planned spending.

The budget's allocation told a clear story about institutional priorities. Aircraft procurement and research, development, test, and evaluation — the line items that fund programs like the F-35, the B-21 Raider stealth bomber, and the Next Generation Air Dominance fighter — consumed approximately $68.3 billion. Shipbuilding claimed $27.2 billion for 17 vessels, anchored by Columbia-class ballistic missile submarines at roughly $9 billion each and Virginia-class attack submarines at approximately $4.3 billion per hull. The Army's ground vehicle modernization budget, including programs to replace the Abrams tank and the Bradley fighting vehicle, added another $12.4 billion.

Unmanned and autonomous systems — the category that includes everything from the Army's small tactical drones to the Air Force's Collaborative Combat Aircraft program to the Navy's unmanned surface vessels — received approximately $4.6 billion.

The ratio was roughly 20 to one. For every dollar the Pentagon spent on unmanned and autonomous systems designed for the current threat environment, it spent twenty dollars on crewed legacy platforms designed for the last one.

That 20:1 ratio was not an accident. It was the mathematical expression of an institution that had spent seven decades building its force structure around exquisite, expensive, human-crewed platforms — aircraft carriers, stealth fighters, main battle tanks — and had not yet restructured its spending to account for the fact that those platforms were becoming progressively cheaper to threaten and progressively more expensive to defend.

The Replicator Gap

The most visible attempt to close the gap was the Replicator initiative. In August 2023, Deputy Secretary of Defense Kathleen Hicks stood at the National Defense Industrial Association's Emerging Technologies conference and announced what she called "a bold new effort" to field autonomous systems at scale. The goal was "multiple thousands" of attritable autonomous systems fielded within 18 to 24 months — by August 2025.

Hicks designed the initiative to bypass the Pentagon's normal procurement timeline, which typically stretched seven to fifteen years from requirement to fielding. Hicks framed the urgency in explicitly economic terms: the United States needed cheap, expendable systems to counter China's advantage in mass production. "We must ensure the PRC leadership wakes up every day, considers the risks of aggression, and concludes, 'Today is not the day,'" Hicks said.

By August 2025, the initiative had fielded hundreds of systems, not the thousands it promised. The specific numbers remained classified, but defense media reporting, Congressional testimony, and DoD officials' own public statements made clear that Replicator had fallen short of its stated objective. The bureaucratic friction that Hicks had sought to bypass — requirements processes, testing protocols, acquisition regulations, service-level resistance — had reasserted itself. Programs that were supposed to move in months moved in quarters. Contracts that were supposed to be awarded in weeks required months of legal review.

Secretary of Defense Pete Hegseth restructured the effort in late 2025, folding it into a new Defense Autonomous Working Group — the DAWG — that consolidated oversight of all autonomous systems programs under a single office. The restructuring acknowledged what defense analysts had been arguing for months. Replicator's problem was not a lack of technology or funding but a structural mismatch between the speed of the threat and the speed of the institution.

The DAWG inherited Replicator's ambitions and its limitations. It also inherited a fundamental question the initiative had never answered. Even if the Pentagon could field thousands of autonomous systems, where would the money come from? The $4.6 billion allocated to unmanned systems was a rounding error in an $892.6 billion budget. Scaling to the tens of billions required would mean cutting something else — and every dollar in the defense budget had a contractor, a Congressional district, and a four-star general defending it.

The Vulnerability Arithmetic

The budget misalignment was not merely wasteful. It was dangerous.

Consider the aircraft carrier. The USS Gerald R. Ford (CVN-78), the Navy's newest supercarrier, cost approximately $13.3 billion to build. Its air wing, when fully equipped with F-35C Lightning IIs and F/A-18E/F Super Hornets, represented another $8 billion to $10 billion in aircraft. Its crew numbered approximately 4,500 sailors. Its escort group — cruisers, destroyers, and submarines — added tens of billions more.

A swarm of FPV drones configured for maritime strike, each costing $35,000 with extended range and maritime guidance, could be assembled for less than the cost of one of the Ford's steam catapults. One hundred such drones would cost $3.5 million. One thousand would cost $35 million — less than the annual maintenance cost of a single F-35C. The drones would not need to sink the carrier. They would need to damage the flight deck, destroy aircraft on deck, or disable the ship's radar and communications arrays. At those numbers, the defender's magazines would empty long before the attacker's inventory.

The same arithmetic applied across the force structure. The F-35A, at $80 million per copy, could theoretically be threatened on the ground by drones costing $500 apiece. The Abrams M1A2 SEPv3 tank, at approximately $10 million per vehicle, was already being destroyed in Ukraine by FPV drones costing less than $500. The Patriot air defense battery, at approximately $1.1 billion per unit with interceptors, had been used to shoot down drones and cruise missiles that cost a fraction of the interceptor price — each PAC-3 MSE interceptor cost approximately $4 million.

The common thread was not that these platforms were obsolete. They were not. An F-35 could do things no drone could do. A carrier strike group projected power in ways that no swarm could replicate. But the economic logic that justified their cost — the assumption that expensive platforms would face expensive threats — had been inverted. The platforms were designed for a world in which the enemy's weapons cost millions of dollars. They now faced a world in which the enemy's weapons cost hundreds.

The Munitions Dilemma

The cost inversion extended beyond platforms to the weapons they carried. The United States and Israel's joint air campaign against Iran in April 2025 consumed an estimated 1,200 precision-guided munitions in a single night — Tomahawk cruise missiles ($2.1 million each), Joint Air-to-Surface Standoff Missiles ($1.4 million each), and Standard Missile-6 interceptors ($4.7 million each) expended against Iranian ballistic missiles, cruise missiles, and one-way attack drones.

The strike achieved its tactical objective. But the munitions expenditure exposed a structural vulnerability: the United States could not replace those weapons quickly. Production lines for Tomahawks and JASSMs ran at peacetime rates — dozens per month, not hundreds. A sustained conflict against a peer adversary would exhaust inventories within weeks. The Congressional Research Service estimated that rebuilding the munitions expended in the Iran campaign alone would take 18 to 24 months at current production rates.

Low-cost alternatives — the kind of cheap, attritable weapons that could be produced at scale and expended without strategic consequence — were years away from operational deployment. The Navy's efforts to develop a long-range, low-cost cruise missile had produced concepts but not production contracts. The Army's precision strike missile program was fielding weapons at $2.5 million per round — an improvement over Tomahawks but still orders of magnitude more expensive than the threats they were designed to counter.

The munitions crisis was the cost inversion expressed in the language of logistics: the United States was burning through irreplaceable complex weapons to counter threats that cost a fraction of the interceptor price, and the production base could not keep pace with expenditure rates in a serious conflict.

The Counter-Drone Billions

The most explicit acknowledgment of the cost inversion in the FY2026 budget was the counter-drone line. The National Defense Authorization Act for fiscal year 2026 authorized $3.187 billion specifically for Counter-Unmanned Aircraft Systems — C-UAS — programs. The DoD's own budget request added a separate $7.5 billion line item for counter-drone capabilities across the services, bringing the total counter-drone investment to more than $10 billion.

Ten billion dollars to defend against threats that cost thousands. The math was a direct descendant of the ABM problem that had forced the Anti-Ballistic Missile Treaty in 1972: when the cost of defense permanently exceeds the cost of offense, the defender's budget becomes the attacker's weapon. Every dollar the Pentagon spent on counter-drone systems was a dollar the attacker had forced it to spend — a dollar diverted from offensive capability, from modernization, from the unmanned systems that could restore the cost advantage.

The counter-drone budget was necessary. No military could ignore the drone threat. But the spending illustrated a deeper structural problem: the Pentagon was paying to defend its expensive platforms from cheap threats while simultaneously buying more expensive platforms that would require more expensive defenses. The cycle was self-reinforcing, and the budget reflected it.

The Institutional Gravity

The 20:1 ratio persisted not because Pentagon leaders were unaware of the cost inversion. They were acutely aware of it. Hicks had articulated the problem precisely when she announced Replicator. The service chiefs had testified to it. The defense media covered it daily.

The ratio persisted because of institutional gravity — the accumulated weight of contracts, constituencies, programs, careers, and Congressional politics that kept the budget anchored to its existing structure. The F-35 program sustained 254,000 jobs across 45 states. The Virginia-class submarine program employed tens of thousands of workers at shipyards in Connecticut and Virginia whose senators sat on the Armed Services Committee. Every major weapons program was simultaneously a military capability, an industrial policy, and a jobs program — and the political cost of cutting any one of them exceeded the political reward of funding its replacement.

Defense analysts at the Center for Strategic and International Studies estimated that shifting even 5% of the total defense budget — roughly $45 billion — from legacy platforms to unmanned systems would require canceling or significantly restructuring at least three major programs of record. Each cancellation would trigger lobbying campaigns, Congressional opposition, and media coverage framed around job losses. The politics of defense budgeting were, in this respect, perfectly designed to prevent the kind of rapid restructuring that the cost inversion demanded.

The result was a budget that acknowledged the drone revolution in its testimony and ignored it in its line items. The Pentagon spent $892.6 billion a year. Roughly $4.6 billion of that — one half of one percent — went to the systems designed for the war that was already being fought. The United States spends twenty dollars on systems designed for the last war for every one dollar it spends on systems designed for this one.

The Budget as Battlefield

The $892 billion question was not whether the United States could afford to restructure its defense budget. It was whether the institution could restructure its defense budget before the cost inversion made the question academic. Every year the 20:1 ratio persisted, the gap between the threat and the response widened. Every year the gap widened, the cost of closing it grew. The budget was not just a spending plan. It was a battlefield, and the United States was losing ground on it at the same rate it was spending money — which was to say, faster than any nation in history.

But the Pentagon's budget was only one side of the ledger. On the other side, a different set of institutions was making a different bet — not on defending expensive platforms, but on building cheap ones. The defense contractors who had spent decades selling the Pentagon its most expensive weapons now had a competitor, and the competitor's business model was built on the opposite premise. That cheap beats expensive, that software beats hardware, and that the first company to build an autonomous weapons platform for $500 instead of $500 million would own the future of defense. The counter-drone industry was where those two bets collided, and the collision was producing a market that looked less like an arms race than a casino.

Chapter 10: The Counter-Drone Casino

The beam was invisible.

On a testing range at the Porton Down defense research facility in Wiltshire, England, in early 2025, a team from MBDA — the European missile consortium — powered up the DragonFire directed-energy weapon for a live engagement against an inbound target drone. The system, developed in partnership with the UK Ministry of Defence, had been in testing for three years. It worked by focusing a high-energy laser beam onto a target with enough precision to burn through the drone's airframe in seconds. The engagement sequence was fast — detection, tracking, lock, fire — and the result was immediate. The target drone disintegrated mid-flight, its composite body superheated beyond structural tolerance. It fell to the ground in pieces.

The assembled officials from the MoD and MBDA applauded. The per-shot cost of the engagement was approximately thirteen dollars — the price of the electricity consumed by the laser and its cooling system during the shot. Thirteen dollars to destroy a drone that, depending on the model, cost anywhere from $500 to $50,000. The cost-exchange ratio was, for the first time in the counter-drone equation, decisively in the defender's favor.

Except that it was not. The DragonFire system had cost approximately $120 million to develop. The laser itself required integration into a naval vessel or ground platform, which meant tapping into a ship's electrical generation capacity, its cooling infrastructure, and its fire-control radar. It could engage one target at a time. In clear weather, with a stable platform, against a single drone flying a predictable approach vector, the system was devastating. Against fifty drones arriving simultaneously from multiple angles in rain or fog or smoke — the actual conditions under which a swarm attack would occur — the system's engagement rate became the bottleneck. Each shot took seconds. Reorienting to the next target took seconds more. The attacker needed only to send more drones than the system could destroy in the available time.

And the attacker could afford to. Fifty FPV drones, at $500 each, cost $25,000. The cost of one of DragonFire's routine maintenance cycles exceeded that figure. The per-shot economics were revolutionary. The system economics were not. DragonFire was a technological triumph and an economic trap — a cheaper way to kill individual drones that still cost more, in total system terms, than the drones it was designed to kill.

The counter-drone industry was selling a solution. The math was selling a stalemate.

The Market Boom

The global counter-drone market did not exist as a distinct category before 2015. By 2025, it was valued at between $4.5 billion and $4.9 billion, depending on the market research firm doing the counting. Projections for the next decade ranged from $14.5 billion by 2030 on the conservative end to $36.4 billion by 2035 on the aggressive end, representing compound annual growth rates of 25% or higher.

The growth was driven by demonstrated threat. Every month brought new footage from Ukraine of drones destroying armored vehicles, ammunition depots, and command posts. Every month brought new reports of drone incursions over military bases in the United States, the United Kingdom, and across Europe — incursions that existing air defense systems were not designed to detect or engage. Every month brought new contract announcements, new startup pitches, new Congressional earmarks.

The market's taxonomy was sprawling. Counter-drone systems fell into four broad categories: detection (finding the drone), tracking (following it), identification (determining whether it is a threat), and defeat (neutralizing it). Defeat systems subdivided further into electronic warfare — jamming the drone's radio link or spoofing its GPS signal — kinetic interceptors, directed-energy weapons, and net-based or entanglement systems. Each category had dozens of competitors, each competitor had a pitch deck, and each pitch deck promised the same thing — a solution to the drone problem.

The Department of Defense was the world's largest buyer. The National Defense Authorization Act for fiscal year 2026 authorized $3.187 billion specifically for C-UAS programs, and the Pentagon's own budget request added $7.5 billion across the services for counter-drone capabilities. The combined investment exceeded $10 billion — more than the entire annual defense budget of most NATO member states. The money was flowing, and the defense industry was building the pipe.

The Jamming Trap

The most widely deployed counter-drone technology was electronic warfare — radio-frequency jamming systems that disrupted the communication link between a drone and its operator, causing the drone to lose control and crash or return to its launch point. Systems like the L3Harris Vampire, the SRC SkyChaser, and the Lithuanian NT Service Group's EDM4S were in active use across multiple theaters.

Jamming worked. Against a standard FPV drone relying on a 5.8 GHz video link and a 2.4 GHz control link, a purpose-built jammer could sever the connection at ranges of several hundred meters to several kilometers. Russian forces in Ukraine had deployed electronic warfare systems at scale, and the effect on Ukrainian drone operations was significant — some units reported loss rates exceeding 60% in heavily jammed sectors.

But jamming had a structural weakness that no amount of investment could resolve: it was a software problem pretending to be a hardware solution. When a jammer defeated a drone by disrupting its radio link, the attacker's response was not to build a more expensive drone. The response was to change the frequency, add frequency-hopping protocols, switch to a different communication method, or — most consequentially — give the drone enough onboard autonomy to complete its mission without a radio link at all.

The fiber-optic drone was the clearest expression of this dynamic. By running a thin fiber-optic cable from the operator's station to the drone, engineers eliminated the radio link entirely. The drone's video feed and control signals traveled through the cable, immune to radio-frequency jamming. Russian forces began deploying fiber-optic FPV drones in significant numbers in late 2024, and by early 2026, Ukrainian military sources estimated that fiber-optic variants constituted a growing share of Russian drone attacks in the most heavily jammed sectors.

A jamming system that cost millions of dollars to develop and deploy could be defeated by a fiber-optic cable that cost a few dollars per meter. The counter-drone industry had invested billions in electronic warfare. The attacker had invested pennies in a spool of glass fiber.

The Kinetic Dilemma

When jamming failed, the next option was to shoot the drone down. This was the oldest solution to the oldest military problem — destroy the threat with a projectile — and it carried the oldest military cost problem. The interceptor cost more than the target.

The most creative kinetic solution on the market in 2026 was the Fortem DroneHunter 5.0, an autonomous counter-drone system that used a tethered net to physically capture hostile drones in flight. The DroneHunter was itself a drone — an AI-guided interceptor that detected, tracked, and netted its target without human intervention. The US Army awarded Fortem Technologies an $18 million contract for the system in 2024, and the DroneHunter was selected to provide counter-drone protection for the 2026 FIFA World Cup in the United States.

The DroneHunter was clever. It was cost-effective relative to a $4 million missile interceptor. It was reusable. But it was one-to-one, with each DroneHunter engaging one target at a time. Against a swarm of fifty drones, the defender needed fifty DroneHunters — or needed to intercept and re-deploy the same units fast enough to keep pace with the incoming wave. The system was an engineering success and a scaling challenge. It worked at the tactical level, for point defense of high-value targets. It did not work at the theater level, where the attacker could send drones by the hundred.

Traditional kinetic interceptors — missiles — faced the same cost problem they had always faced, now amplified. A Patriot

PAC-3 MSE interceptor cost approximately $4 million. A Stinger missile cost approximately $400,000. Even the cheapest guided munitions in the US inventory cost tens of thousands of dollars. Shooting a $4 million missile at a $500 drone was not a military tactic. It was a cash-transfer program from the defender's budget to the attacker's strategic advantage.

The US Army's Integrated Visual Augmentation System program had explored using modified small arms and machine guns with AI-assisted targeting to provide a cheaper kinetic option, but the hit probability against fast-moving small drones at tactically relevant ranges remained low without guided projectiles. Coyote Block 3, a small guided counter-drone missile produced by RTX (formerly Raytheon), cost approximately $100,000 per round — cheaper than a Patriot interceptor by a factor of forty, but still two hundred times the cost of the cheapest target it was designed to destroy.

The Directed-Energy Promise

Directed-energy weapons — lasers and high-powered microwaves — were the defense industry's answer to the cost-exchange problem. If the cost per shot could be reduced to the price of electricity, the defender could fire indefinitely without exhausting a magazine. The economic logic was compelling. The engineering reality was harder.

DragonFire was one of several directed-energy programs racing toward operational deployment. Israel's Iron Beam, a ground-based solid-state laser developed by Rafael Advanced Defense Systems, had completed successful intercept tests against rockets, mortars, and drones by 2024 and was moving toward initial operational capability. The US Army's Directed

Energy-Maneuver Short-Range Air Defense system — DE-MSHORAD — had been tested on Stryker vehicles and demonstrated the ability to destroy small drones at ranges of several kilometers. Lockheed Martin's HELIOS (High Energy Laser with Integrated Optical-dazzler and Surveillance) system had been installed aboard the USS Preble, an Arleigh Burke-class destroyer, for at-sea testing.

Each of these systems delivered on the per-shot promise. The cost of a laser engagement was measured in single-digit or low double-digit dollars. But each system also carried the same structural limitations — power requirements that demanded large platforms or dedicated generators, cooling systems that limited sustained firing rates, atmospheric degradation in rain, fog, dust, or smoke, and single-target engagement that created a rate-of-fire bottleneck against swarms.

The most fundamental limitation was one the industry rarely advertised. Directed-energy weapons were defensive systems optimized for engagements at the speed of light against targets moving at the speed of sound. They were devastating against individual drones. They were progressively less effective as the number of simultaneous targets increased, because the beam had to dwell on each target long enough to destroy it before reorienting to the next. Against a swarm of one hundred drones arriving in a thirty-second window, even a laser that destroyed each target in two seconds could engage only fifteen. The remaining eighty-five would reach their targets.

The House Always Loses

The counter-drone market was growing at 25% annually. It was attracting billions in defense spending and venture capital. It was producing genuinely innovative technologies — lasers that cost thirteen dollars per shot, autonomous interceptors that captured drones in nets, electronic warfare systems that could blanket entire sectors. Every individual technology was an improvement over what existed five years earlier.

But the market operated under a structural constraint that no individual technology could overcome: the attacker iterated faster and cheaper than the defender. When the defender deployed jammers, the attacker switched to fiber-optic cables. When the defender deployed kinetic interceptors, the attacker reduced the drone's radar cross-section and flew it lower. When the defender invested in directed-energy weapons, the attacker could coat the drone in reflective material, send it through smoke, or simply send more drones than the laser could engage.

The analogy was a casino — but one where the house always lost. The counter-drone industry was the house, taking bets from governments willing to spend billions on defense. The attackers were the players, and they had an advantage the house could never match. Their innovation cycle was measured in weeks, their costs were measured in hundreds of dollars, and their iteration did not require Congressional approval, production contracts, or integration testing. A Ukrainian drone workshop could design, build, test, and deploy a new variant in less time than the Pentagon took to process a contract modification.

The defense industry's own analysts understood this. A 2025 RAND Corporation study on counter-drone effectiveness concluded that "the cost-exchange ratio remains structurally favorable to the attacker across all currently deployed and near-term projected C-UAS systems," and recommended that the Department of Defense focus investment on "systems whose marginal cost of engagement approaches or falls below the cost of the threat" — which, in practice, meant directed energy and electronic warfare, the two categories with the lowest per-engagement costs but the highest system costs.

The honest assessment was that counter-drone technology was improving. Per-engagement costs were falling. But the attacker's costs were falling faster, and the attacker's ability to adapt was structurally superior to the defender's. The counter-drone industry was not solving the cost inversion. It was managing it — at a price measured in billions of dollars per year, with no endpoint in sight.

The Uncertain Endgame

None of this meant the counter-drone industry was a waste. Military forces needed the capability to defend against drones, and the alternative — no counter-drone defense at all — was untenable. The systems being developed and deployed were saving lives, protecting infrastructure, and imposing costs on attackers who would otherwise operate with impunity.

But the counter-drone market was not, as some of its boosters claimed, a path to solving the $200 war. It was a holding action — an expensive, necessary, and ultimately insufficient response to a threat that grew cheaper and more capable with every passing month. The historical parallel was the torpedo boat destroyer: a vessel purpose-built to counter the torpedo boat, which succeeded in its narrow mission but did not prevent the torpedo from transforming naval warfare. The torpedo boat destroyer became the modern destroyer, a permanently necessary class of warship. The counter-drone system would become a permanently necessary category of defense spending. But neither reversed the revolution it was built to resist.

The Pentagon was spending billions on defense. Venture capitalists were spending billions on offense. The counter-drone industry was trying to split the difference. But none of this addressed the most consequential variable in the equation — the adversary building a drone force designed not to match the counter-drone industry's solutions, but to overwhelm them. That adversary was not iterating from a garage in Ukraine or a startup in Silicon Valley. It was building from the world's largest drone component supply chain, fielding from the world's largest manufacturing base, and spending from a military budget designed to achieve one strategic objective — parity with the United States at a fraction of the cost. The answer to the counter-drone question required looking east.

Chapter 11: The China Math

The aircraft weighed sixteen tons and carried no pilot.

On a flight test facility in Pucheng, Shaanxi Province, December 2025, the vehicle designated Jiu Tian — "Nine Heavens" — lifted off the runway on its maiden flight. It was an ungainly machine, broad-winged and thick-bodied, built for cargo capacity rather than speed. Its fuselage housed a bay large enough to carry 100 smaller drones, each one a self-contained autonomous platform equipped with navigation sensors, a guidance system, and a warhead. The concept was straightforward and old as air warfare itself — a mothership that could deploy its payload at range, saturating an adversary's defenses with more threats than the defense could handle. What was new was the scale, the autonomy, and the cost arithmetic. Each drone in the bay cost hundreds of dollars. The interceptor missiles required to shoot them down cost hundreds of thousands. One Jiu Tian sortie could impose a $50 million defense problem for less than the cost of a single American cruise missile.

The maiden flight lasted 26 minutes. The aircraft climbed, circled the test facility at low altitude, and landed without incident. Chinese state media — the China Central Television Network (CCTV) defense channel and the state-affiliated *Global Times* — published the footage within hours. The narration was measured, the production quality high, the message unmistakable. The People's Republic of China had operationalized the drone mothership concept before the United States, and the PLA intended the world to know it.

At the Pentagon, analysts in the Office of the Secretary of Defense watched the same footage. The Jiu Tian was not a surprise — US intelligence had tracked the program through satellite imagery and open-source Chinese defense publications for at least two years. What was notable was the speed. The program had moved from publicly visible ground testing to maiden flight in under 18 months, a timeline that would have been aggressive for an American program twice as well-funded. The US Air Force's own mothership concept — variously studied under programs at DARPA and the Air Force Research Laboratory — had not advanced beyond paper studies and small-scale demonstrations. The closest American analogue, the XQ-58A Valkyrie Collaborative Combat Aircraft built by Kratos Defense, could carry two small drones in an internal bay, not one hundred.

The Jiu Tian was a signal. And the signal was not about one aircraft. It was about a military establishment that had decided, at every level from the Central Military Commission to the factory floor, that the future of warfare was cheap, autonomous, and mass-produced — and that China intended to own that future.

The Swarm Doctrine

The People's Liberation Army had been publicly experimenting with drone swarms since at least 2017, when the China Electronics Technology Group Corporation (CETC) demonstrated a coordinated flight of 119 fixed-wing drones that autonomously formed shapes, split into sub-groups, and reassembled. The demonstration was choreographed and non-military in its immediate application — closer to a drone light show than a combat operation. But military analysts recognized it as a proof of concept for the coordination algorithms that swarm warfare would require.

By 2023, the demonstrations had escalated. PLA-affiliated research institutions and defense companies had publicly shown swarms of more than 200 coordinated unmanned aerial vehicles in exercises that included simulated target acquisition, coordinated strikes from multiple vectors, and autonomous response to simulated defensive fire. The Department of Defense's annual report on Chinese military power, published in October 2025, assessed that the PLA was "rapidly integrating autonomous and semi-autonomous unmanned systems across all domains — air, maritime, ground, and undersea — with a focus on swarm operations designed to overwhelm adversary defenses through mass and coordination."

The doctrine driving this integration was not ambiguous. PLA strategists had published extensively in Chinese military journals on the concept of "intelligentized warfare" — zhinenghua zhanzheng — a framework that positioned AI and autonomous systems as the decisive factor in future conflict. General Secretary Xi Jinping had personally elevated military-civil fusion (junmin ronghe) as a national strategy, directing the integration of commercial technology — including Shenzhen's massive drone component supply chain — into military applications. The PLA did not need to invent the drone motor, the flight controller, or the lithium-polymer battery. Shenzhen's commercial ecosystem had already done that. The PLA's task was integration: taking the world's cheapest, most abundant drone components and building them into military systems at a scale and speed no other country could match.

The Shenzhen Advantage

The cost advantage was structural, not incidental. The city of Shenzhen and its surrounding Pearl River Delta manufacturing corridor produced the majority of the world's drone components — the brushless motors, electronic speed controllers, flight controllers, cameras, and lithium-polymer batteries that powered every drone on Earth, from consumer toys to military-grade combat systems.

DJI, headquartered in Shenzhen and controlling an estimated 70% of the global consumer drone market as of 2025, was the most visible expression of this ecosystem. But DJI was the tip of a supply chain iceberg. Beneath it were thousands of component manufacturers, many of them small and specialized, that collectively supplied the global drone industry. A brushless motor that cost $3 to produce in Shenzhen cost $15 to produce in the United States or Europe, when comparable domestic manufacturing existed at all.

The PLA benefited from this ecosystem directly. Under China's military-civil fusion framework, the boundary between commercial and military production was deliberately blurred. Companies that manufactured consumer drone components were encouraged — and in some cases directed — to develop military variants. Research funded by PLA-affiliated institutions flowed into commercial companies, and commercial innovations flowed back into military programs. The result was a defense-industrial feedback loop that no Western country could replicate, because no Western country had a comparable concentration of commercial drone manufacturing.

The cost implications were staggering. A combat-configured FPV drone assembled from Shenzhen-sourced components cost the PLA an estimated $150 to $300 per unit at production scale —

roughly 30% to 50% less than the same drone would cost a Ukrainian manufacturer sourcing from the same suppliers, because the PLA could purchase components domestically at factory-gate prices without export logistics or sanctions-evasion costs. At those prices, the PLA could produce ten thousand combat drones for the cost of a single F-35 engine.

The Export Pipeline

China's drone advantage was not limited to its own military. The Aviation Industry Corporation of China (AVIC) and the China Aerospace Science and Technology Corporation (CASC) operated the world's second-largest armed drone export programs, behind only Turkiye's Baykar.

AVIC's Wing Loong series — the Wing Loong I and Wing Loong II — had been exported to Egypt, the United Arab Emirates, Saudi Arabia, Pakistan, Kazakhstan, and at least ten additional countries across Africa and the Middle East. CASC's CH-series, including the CH-4 Rainbow and the CH-5, had been sold to Iraq, Jordan, Myanmar, Nigeria, and several other nations. The CH-4, a medium-altitude, long-endurance armed drone comparable to the American MQ-9 Reaper in role if not in capability, sold for between $4 million and $5 million — roughly half the price of a Reaper and dramatically less than any comparable Western or Israeli system.

The pricing strategy was deliberate. Chinese drone exports undercut American and Israeli competitors by 50% or more, capturing markets in countries that could not afford — or were not permitted to buy — Western alternatives. The United States had historically restricted armed drone exports under the Missile Technology Control Regime (MTCR), limiting sales to close allies.

China, which was not a signatory to the MTCR in the same binding capacity, faced no such constraints. The result was a growing network of Chinese drone clients that created strategic dependencies: countries that operated Chinese drones required Chinese maintenance, Chinese spare parts, Chinese software updates, and Chinese training. Each sale was a ten-year relationship, not a one-time transaction.

The Stockholm International Peace Research Institute (SIPRI) documented that China had exported armed drones to at least 16 countries by 2025, making it the world's most prolific supplier by number of client states. Those clients spanned four continents and included nations on every side of every major regional conflict. China had sold armed drones to both sides of the Libyan civil war, to rival powers in the Persian Gulf, and to nations in sub-Saharan Africa that had never previously possessed precision strike capability. The export program was simultaneously a revenue source, a strategic influence tool, and a real-world testing ground for Chinese drone technology.

The AI Race

The competition between the United States and China for AI military superiority was, in most public discussions, framed as a question of who had the best artificial intelligence. American commentators pointed to the dominance of US companies in frontier AI research — OpenAI, Anthropic, Google DeepMind, Meta — and to the United States' lead in large language models, foundation models, and the semiconductor design that powered them. The National Security Commission on Artificial Intelligence, chaired by former Google CEO Eric Schmidt, had warned in its 2021 final report that "China is on pace to be the world's AI leader by 2030" — but also acknowledged that the United States retained significant advantages in AI talent, research, and the most advanced model architectures.

That framing missed the point. The AI military balance was not determined by who had the most sophisticated model. It was determined by who could field AI-enabled autonomous systems at scale, at speed, and at the lowest cost. And on that metric, the competition looked different.

China's advantage was not in frontier AI research. It was in integration and deployment. The PLA's ability to take a commercially developed computer vision algorithm, embed it in a $300 drone, test it in a military exercise, iterate on the results, and field the updated version across thousands of platforms in months was structurally superior to the Pentagon's ability to do the same thing. The American defense acquisition system — even after reforms under DIU, Replicator, and the DAWG — operated on timelines measured in years. The Chinese system, operating under military-civil fusion with direct access to Shenzhen's manufacturing base, operated on timelines measured in months.

The comparison was not theoretical. In 2024, the PLA conducted a series of exercises in the South China Sea that incorporated, according to DoD assessments and satellite imagery analysis by the Center for Strategic and International Studies' Asia Maritime Transparency Initiative, more than 200 unmanned platforms operating in coordinated formations across air, surface, and subsurface domains. The US Navy's unmanned surface vessel program, by comparison, had fielded fewer than a dozen experimental platforms for testing by the same period.

The gap was not in ambition. The Pentagon's strategy documents — the National Defense Strategy, the Joint Warfighting

Concept, the service-specific modernization plans — all called for dramatically increased investment in unmanned and autonomous systems. The gap was in execution: the speed at which strategy became hardware, and the cost at which hardware became mass.

The $3 Motor Problem

The competition reduced, at its core, to a single component: the brushless motor.

A brushless motor of the type used in military-grade FPV drones cost approximately $3 to produce in Shenzhen at scale. The same motor, if manufactured in the United States from domestically sourced materials and components, would cost $12 to $18 — assuming a domestic production line existed, which in most cases it did not. The price difference was a function of labor costs, supply chain density, raw material access, and decades of manufacturing ecosystem development that could not be replicated by policy fiat.

The motor was the smallest component in the drone and the largest symbol of the competition's underlying economics. The United States could build better AI models. It could design more sophisticated autonomous systems. It could write doctrine that was more nuanced and field weapons that were more precise. But if every motor, every battery, every flight controller, and every electronic speed controller in those systems was manufactured in China or from Chinese-sourced components, the supply chain vulnerability was existential. And if the PLA could build the same systems for one-third the cost, the cost-exchange ratio that defined the $200 war applied not only to drones versus tanks but to the Chinese military versus the American one.

The Department of Commerce had attempted to address the supply chain risk through export controls on advanced semiconductors and manufacturing equipment, beginning with the October 2022 restrictions and expanded in subsequent rounds. Those controls targeted the high end of the chip market — the AI accelerators and advanced logic chips that powered frontier models. They did not address the low end — the microcontrollers, motor drivers, and sensor chips that powered drones. A $3 motor did not require a $40,000 Nvidia GPU. It required a $0.50 microcontroller, and China manufactured billions of those annually.

The Strategic Calculus

China did not need to build a better fighter jet than the F-35. It needed to build a thousand drones that cost less than one of the F-35's tires. It was already doing both.

The AVIC J-20 Mighty Dragon, China's fifth-generation stealth fighter, was an impressive aircraft — a legitimate peer competitor to the F-35 in several performance categories, produced at a fraction of the cost. But the J-20 was the conventional dimension of the competition. The unconventional dimension — the dimension that defined the $200 war — was the PLA's ability to complement that aircraft with thousands of autonomous drones at a cost the United States could not match through spending alone.

The US-China military balance in 2026 was not a question of which country had the best weapons. It was a question of which country could field the most weapons at the lowest cost, iterate the fastest, and sustain production at scale in a protracted conflict. On every one of those metrics, China's structural advantages — the supply chain, the manufacturing base, the military-civil fusion doctrine, the cost of production — gave it a position that American defense spending, at any level, could not simply buy its way past.

The $892 billion defense budget was not small. It was the largest in the world by a factor of three, exceeding China's official defense budget of approximately $231 billion. But raw spending comparisons were misleading. China's defense dollar bought more than America's defense dollar, because Chinese labor costs, manufacturing costs, and procurement costs were lower. The International Institute for Strategic Studies estimated that China's defense spending, adjusted for purchasing power parity, was equivalent to roughly $470 billion — still less than the US budget, but far closer than the nominal figures suggested.

And the drone competition inverted the calculus further. In a conflict defined by cheap autonomous mass, the country that could produce ten drones for every one its adversary fielded held an advantage that no amount of per-unit superiority could overcome. The United States was building the best autonomous systems money could buy. China was building the most autonomous systems money could buy. In the $200 war, quantity had a quality all its own.

The Race That Cannot Be Won by Spending

The United States could not win the drone competition by outspending China on expensive platforms. The economics did not allow it. An $892 billion budget structured around $170 million fighters, $9 billion submarines, and $13 billion aircraft carriers was optimized for a competition defined by unit cost and individual capability. The drone competition was defined by unit cost and aggregate quantity. The country that built the $3 motor would set the cost floor, and the country that set the cost floor would determine the exchange rate of the conflict.

The path to competitive advantage required something the Pentagon had so far refused to do — restructure the defense budget around cheap autonomous mass. Not as a supplement to existing programs, but as a replacement for a significant share of them. Not $4.6 billion in unmanned systems alongside $68 billion in manned aircraft, but a fundamental rebalancing that matched investment to threat.

The counter-drone industry was spending billions on defense. The Pentagon was spending hundreds of billions on platforms. China was spending fractions of both to build the force that could overwhelm them. The math was not ambiguous. The institutional resistance to acting on the math was.

But the competition with China, as consequential as it was, addressed only one dimension of the $200 war — the great-power dimension, where nations with enormous budgets competed for advantage. The most dangerous dimension was not the one defined by the country that spent $892 billion or the country that built the $3 motor. It was the one defined by what happened when the weapons became so cheap that the budget did not matter at all — when any country, any group, anyone with a wire transfer could afford the same capability that great powers were spending billions to develop. That dimension required looking not at who was winning the drone race, but at who had just entered it.

Chapter 12: The Drone Bazaar

Sergeant First Class Amadou Diallo had never seen a Bayraktar TB2 in person before the morning of April 14, 2024. He had seen the footage, of course — everyone in the Nigerien military had watched the Azerbaijan videos, the grainy thermal clips from Libya, the Turkish propaganda reels set to orchestral music. But standing on the tarmac at Niamey's Diori Hamani International Airport as a Turkish Air Force cargo transport lowered its ramp, Diallo confronted the physical reality of a revolution his country had purchased for less than the cost of a single used F-16. The crated TB2 on the pallet weighed about 650 kilograms. It could stay aloft for 27 hours. It carried four MAM-L laser-guided munitions, each capable of destroying a vehicle or a fortified position from 25,000 feet. Niger's entire previous air force consisted of two aging Sukhoi Su-25 attack aircraft and a handful of transport helicopters. As of that morning, it possessed a precision strike capability that a decade earlier would have required a $3 billion investment in manned fighter aircraft, pilot training, maintenance infrastructure, and logistics chains. The TB2 cost between $2 million and $5 million, depending on the configuration and the contract terms Ankara offered its clients.

Diallo was not thinking about geopolitics. He was thinking about the Sahel, about the armed groups that had turned the border regions into ungoverned corridors of violence, about the fact that his country now had eyes — and weapons — that could reach them from five miles up. But the transaction he witnessed that morning was part of something far larger than Niger's security crisis. It was one node in a global proliferation network that had, in less than a decade, shattered the precision strike monopoly that the United States, Israel, and a handful of NATO allies had maintained since the first Gulf War.

Turkey's Drone Empire

The numbers tell the story with brutal clarity. By 2024, Turkiye had captured roughly 65% of the global armed drone export market, a dominance unmatched by any single country in any major weapons category since the United States dominated fighter jet exports during the Cold War. Baykar, the privately held Turkish company that manufactures the TB2 and its larger successor, the Akinci, had signed export contracts worth $2.2 billion with more than 20 countries spanning four continents.

The client list reads like a map of the world's most volatile regions. Ukraine, which deployed TB2s to devastating effect in the opening weeks of Russia's 2022 invasion. Azerbaijan, which used them to systematically dismantle Armenia's Russian-supplied air defenses during the 2020 Nagorno-Karabakh war. Ethiopia, which deployed them against Tigrayan forces. Libya, where they helped turn the tide in the country's civil war. Poland, the first NATO member to purchase the system. And across Africa — Niger, Nigeria, Morocco, Tunisia, Togo, and others — where the TB2 arrived as the most consequential military import since the AK-47.

The economics of Turkey's drone diplomacy are elegantly simple. A fourth-generation fighter jet — an F-16, a Rafale, a Eurofighter Typhoon — costs between $50 million and $170 million per airframe, requires years of pilot training, demands a sophisticated maintenance infrastructure, and comes with strings attached: end-user agreements, Congressional notifications, diplomatic leverage. A TB2 costs roughly $2 million to $5 million, can be operated by a small team trained in months, and Ankara sells it without the political conditions that Washington or Paris attach to major arms deals.

The result is a new kind of arms market — one where precision strike capability is priced at the level of a luxury apartment, not a warship.

The African Transformation

Nowhere has this market reshaped the security landscape more dramatically than in Africa. In 2024, the Africa Center for Strategic Studies and the Armed Conflict Location and Event Data Project (ACLED) documented 484 drone strikes across 13 African countries — a figure that would have been unthinkable five years earlier. Sudan and the Sahel accounted for 84% of those strikes. In Sudan's civil war, both the Sudanese Armed Forces and the Rapid Support Forces deployed armed drones, some supplied by Iran, others by the United Arab Emirates through intermediary channels.

The pattern was consistent. Countries that had been excluded from precision warfare for decades — not for lack of desire but for lack of budget — were acquiring strike capability at consumer-grade prices. The TB2 was the gateway product, but it was not alone.

The strategic implications extended beyond individual strikes. For the first time in the post-colonial era, African militaries possessed organic precision strike capability that did not require an American or European patron to sustain. The logistics model that sustained F-16 operations in Egypt or Saudi Arabia — dependent on US contractor support, Congressional approval for spare parts, and a multi-billion-dollar maintenance infrastructure — had no analog in the TB2 ecosystem. Baykar trained local crews, supplied spare parts directly, and operated on contract terms that did not require the buyer to align its foreign policy with Ankara's preferences. The drone had severed the link between precision strike and great-power dependency.

China's Parallel Pipeline

While Turkey dominated the export market in volume, China built a parallel pipeline that operated on different terms. The Aviation Industry Corporation of China (AVIC) sold its Wing Loong I and Wing Loong II armed drones to Egypt, the United Arab Emirates, Saudi Arabia, Pakistan, Kazakhstan, and at least ten additional countries across Africa and the Middle East. The China Aerospace Science and Technology Corporation (CASC) exported its CH-4 and CH-5 series on similar terms. The pricing was aggressive: Chinese drones typically undercut American and Israeli equivalents by 50% or more, and the contracts came with no human rights conditions, no end-user monitoring, and no Congressional notifications.

By 2025, China had sold armed drones to 16 or more countries, according to data compiled by the Center for a New American Security (CNAS) drone proliferation dataset and the Stockholm International Peace Research Institute (SIPRI). The sales were not just transactions. They were infrastructure investments. Chinese drone exports came bundled with training, maintenance contracts, and — critically — dependency on the Shenzhen supply chain for replacement parts and software updates. Every Wing Loong sale was a long-term commercial relationship that bound the buyer to Beijing's industrial ecosystem.

The strategic logic was not subtle. The United States had spent decades using fighter jet exports as a tool of alliance management — selling F-16s to allies, withholding them from adversaries, using maintenance contracts and spare parts as levers of influence. China was replicating the model with drones, at a fraction of the price, with none of the political constraints.

The competitive dynamics between Turkey and China were reshaping the global arms market along a new axis. Traditional arms exporters — the United States, France, Russia — sold platforms that locked buyers into long-term dependency relationships through training, maintenance, and spare parts. Turkey and China sold platforms that undercut the incumbents on price and terms, capturing market share in regions where the traditional exporters had either priced themselves out or imposed conditions that buyers found unacceptable. For countries in Africa, the Middle East, and Central Asia that wanted precision strike capability without the political entanglements of a US or European arms relationship, the drone bazaar offered exactly what the legacy market could not — affordable capability, delivered fast, with no questions asked.

Iran's Garage-Scale Proliferation

Turkey sold state-level platforms. China sold state-level platforms at a discount. Iran did something different. Iran sold disposable weapons to non-state actors at prices that made the TB2 look extravagant.

The Shahed series of one-way attack drones — particularly the Shahed-136, the delta-winged system that became a fixture of Russia's bombardment of Ukrainian cities — cost an estimated $70,000 to $80,000 per unit in Russian domestic production, after

Iran originally sold the weapon to Russia at approximately $193,000 per unit. Iran shipped approximately 600 disassembled Shahed-136 units to Russia, along with components sufficient to assemble an additional 1,300 or more, according to US and European intelligence assessments made public through sanctions documentation and Congressional briefings.

But Russia was only the highest-profile customer. Iran supplied drone technology and complete systems to Houthi forces in Yemen, Hezbollah in Lebanon, and Popular Mobilization Forces (PMF) in Iraq. The Houthi drone and missile campaign against commercial shipping in the Red Sea was built substantially on Iranian-supplied technology.

Iran's proliferation model was distinct from Turkey's or China's in one critical respect: it was designed for proxy warfare. The Shahed was not a prestige platform meant to anchor a national air force. It was a disposable weapon meant to impose costs on adversaries through sheer volume. At $70,000 to $80,000 per unit in volume production, a non-state actor could procure a dozen or more Shaheds for the price of a single TB2 — and the Shahed did not require an airfield, a maintenance crew, or a ground control station. It required a launcher, a target, and a willingness to use it.

The three-tier structure of the drone bazaar — Turkey at the top with state-level platforms, China in the middle with discounted alternatives, Iran at the bottom with disposable proxy weapons — meant that precision strike capability was now available at every price point, to every category of buyer, through supply chains that operated independently of each other and could not be shut down by any single regulatory action. A nation that could not afford a TB2 could buy a Wing Loong. A non-state group that could not buy a Wing Loong could acquire Shaheds. And below the Shahed, the $200 FPV drone waited — requiring no state supplier at all.

The Proliferation Staircase

The aggregate numbers form a staircase that only climbs. In 2010, three countries possessed armed drones: the United States, Israel, and the United Kingdom. By 2020, the number had grown to roughly 20. By 2024, more than 40 countries operated armed drone systems of some kind, from the TB2 to the Wing Loong to the Shahed to domestically produced platforms in countries like Iran, South Korea, and India.

Turkish arms exports to Africa alone grew over 700% in a single year — 2021 — driven almost entirely by drone sales in the wake of Azerbaijan's military victory over Armenia. That war, fought over six weeks in the fall of 2020, was the proof of concept that broke the market open. Azerbaijan deployed a combination of Turkish TB2s and Israeli-made HAROP loitering munitions to systematically destroy Armenian air defenses, armored vehicles, and artillery positions. The footage — released in near-real time on social media — was watched by every military planning staff on Earth. Within 12 months, Baykar's order book had exploded.

The lesson that military leaders drew from Nagorno-Karabakh was not complicated. A country with a modest defense budget and no history of aerospace manufacturing had, through a combination of purchased drone capability and effective tactical employment, routed a military equipped with Russian-supplied air defenses, tanks, and artillery. The TB2 had done what a fleet of manned aircraft would have done — at a fraction of the cost, with zero pilot casualties.

The footage from Nagorno-Karabakh did something that no defense white paper or think tank report could accomplish. It made the cost inversion visible. Military planners did not read about the TB2's effectiveness in a classified brief. They watched it, on their phones, in thermal video that showed Russian-made Osa and Strela air defense systems exploding under laser-guided munitions dropped from an altitude their operators never detected. The visual evidence was more persuasive than any procurement argument. Baykar's export pipeline, which had been a modest stream before the 2020 war, became a torrent within the year. Countries that had been negotiating for years suddenly accelerated their timelines. Countries that had not considered drone procurement at all began inquiries. The Nagorno-Karabakh footage was the advertisement, and the entire developing world was the market. The precision strike monopoly was not eroding. It was gone.

The Next Wave

The proliferation pipeline was not slowing. It was accelerating. In 2025, Baykar unveiled the K2, a kamikaze drone with an 800-kilogram maximum takeoff weight, a 200-kilogram warhead, a range exceeding 2,000 kilometers, and AI-assisted swarm capability. The K2 was explicitly designed for export, priced to undercut Western alternatives, and marketed as a system that could operate in coordinated swarms without requiring a dedicated operator for each airframe.

The K2 represented the next step in the proliferation staircase: not just the democratization of precision strike, but the democratization of autonomous, swarming precision strike. A country that bought a K2 package was not just buying individual weapons. It was buying the ability to saturate a target area with coordinated, semi-autonomous munitions — a capability that, five years earlier, only the United States and Israel had demonstrated.

The Bayraktar had given any country an air force. The K2 promised to give any country a swarm.

But the TB2 cost $2 million. The FPV drone — the weapon that had actually killed the most armor, destroyed the most positions, and reshaped the most tactics on the Ukrainian front — cost $200. The gap between a state-level weapon and a garage-level weapon was collapsing at a speed that no export control regime, no arms trade treaty, and no nonproliferation framework had been designed to address. When a precision-guided munition could be assembled from commercial parts for the price of a pair of running shoes, the question was no longer who had access to military-grade strike capability. The question was who did not.

Chapter 13: The $200 Air Force

Olena Zhuravel spread the components across the workbench in her garage workshop outside Dnipro, Ukraine, on a Wednesday evening in October 2025. She was 34 years old, a former IT project manager, and she had built her 412th FPV combat drone that month. The inventory was unremarkable: a carbon fiber frame she bought in bulk from a Chinese supplier through AliExpress for $20 per unit, four brushless motors at $3 each, a flight controller for $8, an analog video transmitter for $10, a set of lithium-polymer batteries for $15, and a 3D-printed tail stabilizer that her desktop printer produced in 22 minutes from a template shared on a volunteer Telegram channel. Total cost of the airframe, excluding the warhead, was $165.

Zhuravel was not a defense contractor. She was one of more than 500 small-scale drone manufacturers who had emerged across Ukraine since Russia's full-scale invasion in February 2022 — a decentralized, garage-level production ecosystem that had collectively manufactured more than 4 million drones in 2024 alone. Her workshop occupied the same space where she had once parked her Skoda. The workbench was a repurposed kitchen table. The soldering station had been a birthday gift from her husband, who was serving with the 93rd Mechanized Brigade somewhere in the Donetsk sector. She tested each drone in the empty lot behind the building, running it through a four-minute flight profile that checked motor response, video link stability, and GPS lock. Then she packed them in foam-lined cardboard boxes, 12 to a case, and drove them to a logistics hub where volunteers sorted shipments by front-line sector.

No export license covered the transaction. No government office approved the components. No international treaty regulated the assembly. Every part on Zhuravel's workbench was commercially available to anyone on Earth with an internet connection and a shipping address.

The Bill of Materials

The economics of the FPV combat drone are so simple they fit on the back of a receipt. A complete airframe — frame, motors, flight controller, video transmitter, receiver, batteries, and propellers — costs between $150 and $500, depending on the components and the quality. The low end of that range, around $200, produces a drone capable of carrying a warhead weighing between 1 and 3 kilograms at speeds exceeding 100 kilometers per hour, with a range of 5 to 15 kilometers depending on the video link. At the high end, $500 buys better cameras, longer range, and more reliable components — but the destructive capability is functionally identical.

The warhead is the one component that does not come from Amazon. In Ukraine, drone workshops typically source warheads from conventional munitions — repurposed RPG-7 rounds, adapted mortar shells, shaped-charge warheads machined in small workshops from readily available materials. The warhead cost varies, but Ukrainian sources consistently report total weapon costs — airframe plus warhead — in the $300 to $500 range for a standard anti-armor FPV drone.

A weapon in that price range, guided by a skilled operator wearing $40 FPV goggles, has destroyed Russian T-72 tanks valued at $1 million to $3 million, BMP infantry fighting vehicles, self-propelled artillery systems, electronic warfare stations, and —

in the case that opened this book — a Tor-M2 surface-to-air missile system valued at $24 million. The cost-exchange ratio is not a rounding error. It is a structural inversion of the economics of warfare.

The bill of materials reveals something else: the weapon has no signature component. There is no single part that distinguishes a combat FPV drone from a racing FPV drone or a photography FPV drone. The flight controller that stabilizes a drone carrying a shaped-charge warhead is identical to the flight controller that stabilizes a drone carrying a GoPro. The motors, the propellers, the frame, the batteries — all of them are dual-use by nature, civilian by origin, and military only by assembly. This is what makes the FPV combat drone fundamentally different from every previous precision weapon. It is not manufactured as a weapon. It is assembled into one.

Five Hundred Factories

Before Russia's invasion, Ukraine had approximately seven companies producing military drones. By mid-2025, the number exceeded 500. The Ukrainian government's Brave1 defense technology accelerator, launched in 2023, had processed more than 2,000 applications from drone developers and funded hundreds of projects. The government allocated $2.5 billion to drone procurement in 2024, a figure that represented both the scale of the need and the recognition that decentralized production was the only model that could meet it.

The production ecosystem operated at every scale simultaneously. At one end, established companies like Ukrspecsystems and Aerorozvidka manufactured sophisticated reconnaissance and strike platforms with advanced avionics and encrypted data links. At the other end, individual volunteers like Zhuravel built basic FPV drones in their garages, funded by donations, crowdfunding campaigns, and their own savings.

The speed of this transformation defied every assumption about defense production. Traditional military procurement operates on timelines measured in years — requirements definition, competitive bidding, contract negotiation, prototype testing, low-rate initial production, full-rate production. Ukraine's decentralized model operated on timelines measured in days. A front-line unit identified a need. A Telegram message reached a workshop. A drone tailored to the requirement was assembled and shipped within the week. The feedback loop between the battlefield and the factory was so short that design iterations happened in real time — operators sent video of failed attacks, and engineers modified the next batch of drones to compensate.

The crowdfunding dimension was itself unprecedented. The Ukrainian volunteer campaign to build Sea Baby uncrewed surface vessels — the sea drones that struck Russian naval targets in the Black Sea — raised $7.8 million in 36 hours. The campaign was not organized by the military. It was organized by civilians on social media, funded by individual donations averaging less than $100, and executed by a network of engineers, fabricators, and boat builders operating outside the formal defense procurement system.

This was the model that no defense ministry in the world had planned for: a nation at war, producing precision munitions at garage scale, funded by crowdfunding, distributed across hundreds of independent workshops, and outproducing its adversary's ability to defend against the weapons it was building.

The production numbers were staggering in their own right. Ukraine manufactured more than 4 million drones in 2024 — a figure that exceeds the total number of guided munitions produced by the entire NATO alliance in the same period. The majority were simple FPV airframes, not sophisticated multi-sensor platforms, but the aggregate effect was transformative. Each drone that destroyed a $1 million tank or a $500,000 armored vehicle represented a net extraction of value from the Russian military at a ratio that no conventional arms production calculus could match. Ukraine was not outspending Russia. It was out-mathing Russia — producing weapons so cheap and so numerous that the cost of defending against them exceeded the cost of producing them by orders of magnitude.

The Swarm Goes Platform-Agnostic

The next step in the democratization arrived in January 2026. Auterion, a Swiss-American drone software company, and Nemyx, a defense AI startup, conducted the first publicly documented live-fire demonstration of a multi-manufacturer swarm. A single operator engaged three separate targets simultaneously using drones built by different manufacturers, running different hardware, unified by a common software layer.

The significance was not the number of targets or the number of drones. The significance was what the demonstration proved about the architecture of cheap warfare. Until that point, swarm operations had been platform-specific — a given swarm required drones from the same manufacturer, running the same firmware, communicating through the same protocols. The Auterion/Nemyx demonstration broke that constraint. It showed that swarm capability was becoming a software problem, not a hardware problem. Any drone that could run the coordination software — regardless of who built the airframe, what motors it used, or what country it came from — could join the swarm.

For a military planner, the implication was immediate. A nation did not need a single drone manufacturer to field a swarm. It needed software. The drones could come from anywhere — from 500 Ukrainian garages, from Shenzhen factories, from Turkish export contracts, from a 3D printer in a basement. The swarm was no longer defined by the platform. It was defined by the code.

The January 2026 demonstration marked a threshold that defense analysts had predicted but never verified in live fire — the moment when swarm capability decoupled from hardware. Before the Auterion/Nemyx test, fielding a drone swarm required a vertically integrated system — one manufacturer, one software stack, one communication protocol. After the test, swarm capability became modular. Open-architecture software could coordinate any platform that met a minimum set of interface standards, the same way an operating system can run on hardware from multiple manufacturers. The defense-industrial implications were profound. The barriers to fielding a swarm had just collapsed from "build a complete system" to "install an app."

The Arms Control Void

Every major arms control framework in existence was designed for a world that no longer exists. The Missile Technology Control Regime (MTCR), established in 1987, restricts the transfer of missiles and unmanned aerial vehicles capable of carrying a 500-kilogram payload to a range of 300 kilometers or more. An FPV combat drone carries a 1-to-3-kilogram payload to a range of 5 to 15 kilometers. It does not meet the MTCR threshold by any definition.

The Wassenaar Arrangement on Export Controls for Conventional Arms and Dual-Use Goods and Technologies covers a broader range of systems, but its enforcement depends on the participating states' willingness to restrict exports — and it does not cover commercial components sold through civilian e-commerce channels. A flight controller purchased on AliExpress is not a controlled item under any existing export regime. Neither is a brushless motor, a lithium-polymer battery, a video transmitter, or a carbon fiber tube. Each component has legitimate civilian applications — recreational drones, agricultural sprayers, photography platforms, racing quadcopters. The fact that the same components, assembled in a different configuration, produce a weapon capable of destroying a tank is a problem that no treaty negotiated before 2020 was designed to address.

The Conventional Forces in Europe (CFE) Treaty, the Arms Trade Treaty (ATT), the UN Programme of Action on Small Arms and Light Weapons — none of these instruments contemplates a weapon system assembled from consumer electronics that are sold in every country on Earth.

The fundamental problem is not a gap in the treaties. It is a category error. Arms control was designed to regulate the transfer of identifiable weapons platforms between identifiable state actors through identifiable channels. The FPV combat drone is not transferred. It is assembled. Its components are not weapons. They are electronics. Its supply chain is not a defense-industrial pipeline. It is Amazon, AliExpress, and Alibaba.

The Irreversibility Problem

The democratization of lethal precision strike is not a policy failure that better regulation can fix. It is a structural condition created by the convergence of three forces that cannot be reversed: the miniaturization of electronics, the globalization of commercial supply chains, and the open-source distribution of assembly knowledge.

The miniaturization means that a weapon-grade flight controller fits in the palm of a hand and costs less than a restaurant meal. The globalization means that the components are manufactured in dozens of countries and sold through platforms that ship to every address on Earth. The open-source knowledge means that the assembly instructions — wiring diagrams, firmware configurations, warhead integration guides — circulate on Telegram channels, YouTube tutorials, and defense forums in a dozen languages.

No embargo can intercept a supply chain that runs through consumer electronics retailers. No export control can restrict a component that is also used in agricultural drones, delivery drones, and children's toys. No classification regime can protect information that is freely shared by thousands of independent developers, combat veterans, and volunteer engineers.

The implications extend beyond state failure and terrorism. The same $200 weapon that Ukraine built to defend against Russia's invasion could be assembled by a cartel, a separatist movement, an extremist cell, or a lone individual with technical competence and malicious intent. The threat is not speculative. Drug trafficking organizations in Mexico have already deployed commercial drones modified to carry explosive payloads. The Islamic State used commercial drones with modified munitions in Mosul in 2017. The technical barrier between those crude early efforts and the precision capability demonstrated daily in Ukraine is not a matter of classified technology. It is a matter of publicly available firmware, a YouTube tutorial, and $200 in components.

Two hundred dollars. A few hours. A garage. That is the new price of a precision air force. No treaty written before 2020 anticipated a weapon assembled from commercial parts available on any e-commerce platform on Earth.

The Bayraktar cost $2 million and gave countries an air force. The FPV drone cost $200 and gave individuals a guided munition. But both the Bayraktar and the FPV drone still required a human at the controls — a pilot watching a screen, an operator wearing goggles, a person making the decision to strike. The human was the last expensive component in the system. Training cost money. Attention cost time. Fatigue cost accuracy. The next step in the cost curve was not a cheaper drone or a cheaper warhead. It was the elimination of the human operator entirely — the moment when AI handled navigation, targeting, and strike without a person in the loop, and the cost of destruction became the cost of the drone and nothing else.

Chapter 14: The AI Cliff

The video was 47 seconds long. It appeared on a Ukrainian military Telegram channel on March 11, 2024, without commentary — just a timestamp, a unit designation, and a brief caption in Ukrainian that translated roughly to "autonomous completion." Within hours, it had migrated to Western defense media. Defense One published an analysis on March 14. The War Zone ran a detailed technical breakdown the same week. Defense Express, the Ukrainian defense outlet that had become the world's most reliable source for frontline drone footage, confirmed the video's provenance through its network of unit-level contacts.

The footage showed a standard Ukrainian FPV drone — commercial components, estimated cost under $500 — in terminal approach toward a Russian military vehicle. Midway through the dive, the operator's first-person video feed dissolved into static. A Russian electronic warfare system, likely an R-330Zh Zhitel or a Pole-21 jammer, had severed the radio link between the drone and its operator. In previous months, across hundreds of similar engagements, this would have ended the sortie. The drone would have lost guidance, spiraled, and crashed into a field.

This drone did not crash. The video, reconstructed from the drone's onboard recording recovered after the strike, showed the aircraft continuing its descent. It adjusted its flight path — a slight lateral correction, visible in the footage as a smooth, unhurried shift in the camera frame. It maintained course. It struck the target. The operator's screen had been black for the final eight seconds of the engagement. The drone completed the strike on its own.

Colonel Yuriy Shchyhol, who oversaw electronic warfare countermeasures for a sector of the eastern front, told Defense Express that the footage represented a capability his command had first observed in late 2023 but that had become increasingly common by early 2024. The drones were running onboard software — loaded before launch, not requiring a data link — that allowed them to maintain course toward a designated target using visual navigation when the operator's control signal was lost. The technical term was autonomous terminal guidance. The practical term, among the operators Shchyhol worked with, was simpler. They called it "flying blind."

The AI that made this possible was not a classified military program. It was software, developed by a combination of Ukrainian defense-tech startups and volunteer engineers, adapted from the same computer vision algorithms that commercial drones used for obstacle avoidance and automated landing. The cost of adding this capability to an existing FPV drone was negligible — a software update, loaded onto the same flight controller that cost $8.

The Jammer-Autonomy Ratchet

The March 2024 footage was not an anomaly. It was a data point on a curve that had been steepening since the first months of the war. The pattern was mechanical, almost predictable: every time Russian electronic warfare capability improved, Ukrainian drones had to become more autonomous to survive.

The dynamic operated like a ratchet. Each turn tightened in one direction. Russia deployed a new jamming system, and Ukrainian drones lost their control links and crashed. Ukrainian engineers responded by adding onboard navigation software that could maintain course through a jamming envelope. Russia expanded the jamming spectrum, and the software had to become more sophisticated — shifting from simple inertial guidance to visual terrain matching to AI-assisted target recognition. Each cycle of electronic warfare improvement forced a corresponding step toward greater autonomy. The ratchet never turned backward. No one removed autonomy once it had been added. The electronic warfare environment guaranteed that every generation of drone was more autonomous than the last.

By mid-2025, Ukrainian defense officials described a frontline environment where electronic warfare had become so dense that operators routinely lost contact with their drones during the final phase of an engagement. The Russian military had deployed electronic warfare systems at a density of approximately one major jamming platform per 10 kilometers of front line, creating overlapping fields of signal disruption that made reliable human control increasingly difficult. In that environment, the choice was binary — accept the loss of the drone, or give the drone the ability to complete its mission without the operator. The economics made the decision obvious. A $200 drone that could fly itself through a jamming envelope was worth more than a $200 drone that could not. The AI cost nothing extra. The software was free to copy, free to distribute, free to update.

The ratchet had a name among defense analysts who tracked it. They called it the jammer-autonomy spiral.

The spiral was not unique to Ukraine. Every military that studied the conflict recognized the dynamic, and every military that planned for future drone operations built the lesson into its development programs. The US Air Force's research on GPS-denied navigation, Israel's investment in autonomous terminal guidance for loitering munitions, China's publicly demonstrated autonomous swarm coordination — all of these programs reflected the same underlying logic. In contested electromagnetic environments, the human link is the weakest link. The solution is not a stronger link. The solution is no link at all.

The Economics of Zero

The cost arithmetic of the $200 war had, until this point, included one component that was not cheap: the human operator. An FPV drone pilot required training — weeks of simulator time, hours of practice flights, familiarization with the specific tactics and counter-tactics of the front-line sector where the pilot would operate. A skilled operator could guide a drone to a target with lethal precision. An unskilled operator would waste the drone. Training cost time, and time in a war of attrition cost lives.

An operator also occupied a scarce resource — attention. One pilot could operate one drone at a time. A pilot who was flying a drone toward a target in sector A was not available to fly a drone toward a target in sector B. The human operator was a bottleneck — the single point in the kill chain that could not be parallelized by spending more money on hardware.

Autonomous terminal guidance changed the math. If the drone could complete its final approach without the operator, the operator could launch a second drone before the first one struck. If the drone could navigate to the target area autonomously and required the operator only for final target selection, the operator could manage multiple drones in sequence — or, with sufficiently advanced coordination software, simultaneously.

The economic endpoint was visible: a $200 drone with a human pilot cost $200 plus training, salary, attention, and time. A

$200 drone that flew itself cost $200 and nothing else. At scale, the marginal cost of precision strike approached zero. Not theoretically. Arithmetically.

The Market for Autonomous War

The defense industry had already priced this trajectory into its projections. The global market for AI-integrated drone systems — encompassing autonomous navigation, computer vision targeting, and swarm coordination software — was valued at $845 million in 2025, according to Grand View Research. The same analysis projected the market to reach $10.9 billion by 2035, growing at a compound annual growth rate of 29.8%. Nearly $11 billion in annual revenue, growing at almost 30% per year, dedicated to making drones smarter, more autonomous, and less dependent on human operators.

The investment was not speculative. It was a response to demonstrated demand. Every military that had observed the Ukraine conflict understood the jammer-autonomy ratchet. Every defense-tech company that competed for drone contracts understood that the next generation of systems would need to operate through electronic warfare environments that degraded or eliminated human control links. Autonomy was not a feature to be debated. It was a requirement imposed by the electromagnetic environment of the modern battlefield.

DARPA's Autonomous Multi-Domain Adaptive Swarms and Signatures (AMASS) program, funded through its Tactical Technology Office, represented the most ambitious American effort to close the gap between semi-autonomous and fully autonomous drone operations. AMASS funded research into autonomous targeting, swarm coordination without centralized control, and adaptive behavior in contested environments — the precise capabilities that the jammer-autonomy ratchet was forcing into existence on the Ukrainian front, formalized into a research program with the resources of the US defense establishment behind it.

Combat-Proven Autonomy

The gap between research and reality was narrowing from the other direction as well. Swarmer, the Ukrainian drone-swarm company that had gone public on the Nasdaq, had deployed its Styx autonomous coordination system in active combat operations on the eastern front. Styx enabled multiple drones to share targeting data, coordinate approach vectors, and execute simultaneous strikes on a single target or distributed strikes across multiple targets — with varying degrees of human oversight depending on the mission profile and the electronic warfare environment.

Swarmer's SEC filings disclosed that the Styx system had been deployed in more than 100,000 combat missions by mid-2025 — a figure that made it, by a significant margin, the most combat-tested autonomous drone coordination platform in the world. The system had been developed under fire, refined through operational feedback from front-line units, and updated on a cycle measured in days rather than the years typical of US defense acquisition.

The combination of Swarmer's Styx and the broader ecosystem of autonomous navigation software emerging from Ukrainian workshops represented something that defense planners had discussed theoretically for a decade: autonomous weapons systems proven in sustained, high-intensity combat. The theoretical had become operational. The operational had become routine.

The distinction mattered enormously. Before Swarmer, every discussion of autonomous weapons in defense circles was, to some degree, speculative — grounded in engineering projections, laboratory demonstrations, and controlled exercises. After Swarmer, the discussion was empirical. Autonomous multi-drone coordination had been tested not in a DARPA exercise against simulated targets, but against a peer military with electronic warfare, air defenses, and the ability to shoot back. The Styx system had logged its 100,000 missions in the most demanding operational environment on Earth. Its performance data was not theoretical. It was actuarial.

The Voice of Restraint

Not everyone read the trajectory with enthusiasm. Dario Amodei, the chief executive of Anthropic, one of the world's leading AI safety companies, addressed the autonomous weapons question directly in a February 2026 public statement during the Pentagon-Anthropic dispute over autonomous weapons. Amodei, whose company built Claude, one of the most capable frontier AI systems in the world, stated that "frontier AI systems are simply not reliable enough to power fully autonomous weapons" — a statement grounded not in pacifism but in engineering assessment.

Amodei's argument was precise. The problem was not that AI could not guide a drone to a target. The March 2024 footage proved that it could. The problem was reliability at scale, in the messy, unpredictable, morally consequential environment of warfare. AI systems hallucinate. They misclassify. They perform brilliantly in training environments and unpredictably in conditions their training data did not anticipate. A 99% accuracy rate sounds impressive until it is applied to 10,000 strikes, at which point it means 100 strikes on the wrong target.

The Lavender precedent was relevant here — though the Israeli AI targeting system described in Book 1 operated in a different context and at a different scale, it demonstrated the gap between what AI targeting could do technically and what it should do ethically. Lavender flagged 37,000 suspected militants in Gaza for potential targeting, according to +972 Magazine's reporting. The system operated with varying levels of human review, and the consequences of its errors were measured in civilian casualties.

Amodei's position — shared by a significant number of AI researchers, though not by the majority of defense-tech executives — was that the economic incentives driving autonomy forward were real and powerful, but that the technology was not yet trustworthy enough to be given lethal authority without meaningful human oversight. The gap between the economic logic and the engineering reality was what he and other AI safety researchers called the reliability problem.

The defense-tech industry heard Amodei's argument and, for the most part, continued building. The economic incentives were too strong, the competitive pressures too immediate, and the battlefield demand too urgent for a caution rooted in "not yet" to override the logic of "good enough."

The Edge of the Cliff

The AI cliff was not a metaphor for a sudden drop. It was a metaphor for a point of no return — the moment when the economic logic of autonomous weapons became so compelling that no military, no defense ministry, and no non-state actor could afford to stay on the human side of the line. The cliff was not ahead. The first steps over its edge had already been taken, in Ukrainian fields where drones flew blind through jammers, in Swarmer's Styx system coordinating strikes across multiple targets, in DARPA's AMASS program funding the next generation of autonomous swarm behavior.

When the drone no longer needed a pilot, the cost of destruction was the cost of the drone and nothing else. That number was $200 and falling.

The AI cliff was not a future scenario. The first steps had already been taken — in Ukrainian fields, in Shenzhen factories, in Silicon Valley labs. The question was not whether the cost of destruction would approach zero. The economics guaranteed it. The question was whether anyone was willing to look at the numbers and act before the math became irreversible. In a closed briefing room in Washington, someone was about to try.

Chapter 15: The $200 War

Dr. Margaret Kowalski stood at the front of Room 2154 in the Rayburn House Office Building on a Tuesday morning in April 2026, facing 14 members of the House Armed Services Committee's Subcommittee on Cyber, Information Technologies, and Innovation. Kowalski was a senior defense analyst at the Congressional Research Service, and she had been asked to brief the subcommittee on cost-exchange ratios in asymmetric conflict — a phrase bland enough to have attracted no press attention and no public attendance beyond the staffers lining the walls. The room was closed to media. The slide deck was unclassified but marked "For Official Use Only." Kowalski had spent four months compiling the data.

Her first slide was a single table. Two columns. The left column was labeled "Attacker's Cost." The right column was labeled "Defender's Cost." The rows descended through history: Agincourt, 1415 — an English longbow costing a few shillings against a French knight's armor worth a year's income. The torpedo, 1940s — a $10,000 weapon against a $100 million battleship. The IED, 2000s — a $20 explosive device against the $45 billion Joint IED Defeat Organization. The FPV drone, 2024 — a $200 weapon against a $24 million Tor-M2 missile system. The Houthi campaign, 2024 — $2,000 drones triggering $2 million interceptor missiles.

Every row showed the same pattern. The attacker's cost stayed flat or declined. The defender's cost escalated. The ratio between them widened with each generation of conflict.

Kowalski did not editorialize. She let the table sit on the screen for 30 seconds before speaking. Then she said: "The cost-exchange ratio in the current conflict environment is the most extreme in the recorded history of warfare. The trend line has been moving in one direction for six centuries. There is no historical precedent for it reversing."

A committee member — later identified in staff accounts as a senior Republican from a district with a major defense contractor — leaned forward and asked the question Kowalski had prepared for: "What does the math say we should do?"

She advanced to the next slide. Five numbers. Five prescriptions. The room went quiet.

The Five Numbers

The $200 war can be reduced to five figures. Each one is specific, sourced, and uncomfortable.

$200. The cost of the weapon. An FPV combat drone assembled from commercial components — a flight controller, four motors, a carbon fiber frame, a video transmitter, and a lithium-polymer battery — costs between $150 and $500. At the low end, which is where volume production operates, the fully assembled airframe costs approximately $200. Add a repurposed warhead, and the total weapon cost ranges from $300 to $500. This is the price of the weapon that has destroyed more Russian armored vehicles in Ukraine than any missile system in NATO's inventory.

120,000:1. The most extreme cost-exchange ratio in military history. A $200 FPV drone destroying a $24 million Tor-M2 surface-to-air missile system produces a cost-exchange ratio of 120,000 to 1. Even the more common engagement — a $400 drone destroying a $2 million T-72 tank — produces a ratio of 5,000 to 1. For comparison, the IED-to-MRAP ratio that consumed $45 billion in US counter-IED spending was roughly 2,000 to 1. The torpedo-to-battleship ratio in World War II was approximately 10,000 to 1. The FPV drone has exceeded them all.

$892.6 billion. The US defense budget for fiscal year 2026, as authorized by the National Defense Authorization Act. Of that sum, the vast majority funds legacy platforms, force structure, and procurement programs designed for the last generation of warfare. The ratio of spending on legacy manned systems to spending on unmanned and autonomous systems remains approximately 20 to 1. The budget is structured for a $10 million war. The battlefield has moved to a $200 war.

$49.1 billion. The cumulative venture capital investment in defense technology companies through mid-2025, according to PitchBook data. This figure represents the financial market's verdict on where warfare is heading. Anduril, valued at $60 billion as of its March 2026 funding round. Shield AI, valued at $5 billion as of its August 2025 Series G round. Swarmer, trading on the Nasdaq. Skydio, Saronic, Epirus, Fortem Technologies — dozens of companies, funded by billions of dollars, all building toward the same thesis: cheap, autonomous, attritable systems will replace expensive manned platforms. Wall Street has already done the math. The Pentagon has not.

40+. The number of countries that now possess armed drone systems. In 2010, the number was three. The proliferation curve has no historical parallel in the spread of military technology. It took nuclear weapons 25 years to proliferate from one state to five. It took armed drones 15 years to proliferate from three states to more than 40. The precision strike monopoly held by the United

States and its allies for three decades after the Gulf War is gone.

Prescription One: Restructure the Budget

The first prescription follows directly from the 20:1 ratio. The United States cannot fight a $200 war with a budget structured for a $10 million war. The current defense budget allocates roughly 20 dollars to manned legacy platforms — fighters, bombers, surface combatants, armored vehicles — for every dollar it allocates to unmanned and autonomous systems. This ratio must reverse, and the timeline for reversal is not a generation. It is a budget cycle.

The model already exists. The US Army's $52 million contract with Skydio for 2,500 X10D reconnaissance drones, awarded through an Other Transaction Authority (OTA) agreement in 72 hours from solicitation to award, demonstrated that the Pentagon can move at commercial speed when it chooses to. The problem is not capability. It is will. The institutional resistance to shifting funding from legacy programs to attritable systems is not technical. It is political. Every legacy platform has a Congressional district, a prime contractor, a lobbying operation, and a workforce. The FPV drone does not. Restructuring the budget to match the threat environment means dismantling the political economy that sustains the current one. There is no version of this that is painless.

Prescription Two: 72-Hour Acquisition

The second prescription attacks the procurement timeline. The traditional defense acquisition process — from requirements generation through competitive bidding through contract award through production through fielding — takes years. Often a decade. The Skydio X10D contract proved that 72 hours is possible for attritable systems procured through OTA pathways.

The 72-hour timeline must become the standard, not the exception, for all systems that cost less than $50,000 per unit. Attritable drones, loitering munitions, electronic warfare payloads, counter-drone effectors, and autonomous navigation software packages should be procured through streamlined pathways that treat them as consumables, not as programs of record. The Defense Innovation Unit (DIU), established to bridge the gap between Silicon Valley and the Pentagon, has demonstrated the model. Congress must codify it — mandating that all attritable systems below a cost threshold are exempt from traditional acquisition regulations and subject to expedited OTA procurement.

The alternative is a military that fields its response to the $200 drone on the $200 drone's timeline — which means fielding it after the war is already lost.

Prescription Three: Directed Energy at Scale

The third prescription addresses the defender's side of the cost equation. Every analysis in this book has demonstrated that the defender's cost curve rises faster than the attacker's. Interceptor missiles cost orders of magnitude more than the drones they destroy. Kinetic counter-drone systems — bullets, nets, jamming — are effective in limited engagements but cannot scale to meet a saturation attack of hundreds or thousands of cheap drones.

Directed-energy weapons — high-energy lasers and high-powered microwave systems — are the only counter-drone technology that can match the attacker's cost curve. A laser that costs $1 to $10 per shot, powered by a generator running on diesel fuel, can destroy an unlimited number of $200 drones without exhausting a finite magazine. The United Kingdom's DragonFire laser, which completed shipboard trials in 2025, demonstrated a cost-per-shot below $13. Israel's Iron Beam, a ground-based laser designed to complement the Iron Dome, has been deployed in limited operational testing.

Directed energy must move from demonstration programs to deployed infrastructure. The current approach — funding small numbers of prototypes, testing them in controlled environments, and debating their readiness levels — is a luxury the cost curve does not afford. Every month that directed-energy deployment is delayed is a month in which the attacker's advantage compounds. The prescription is specific. Fund the production, not the research. The technology is proven. The engineering challenges are real but solvable. The cost-exchange math demands that directed-energy counter-drone systems be fielded at the scale of the threat, which means thousands of units across every forward-deployed US military installation, every allied base, and every naval vessel in contested waters.

Prescription Four: Lead the Governance Framework

The fourth prescription is the most counterintuitive: the United States should lead the international effort to establish binding norms on autonomous weapons.

In December 2024, 166 nations voted in the UN General Assembly to begin negotiations on a legally binding instrument to regulate lethal autonomous weapons systems (LAWS). The resolution passed with 166 votes in favor, with the United States abstaining and only Belarus, North Korea, and Russia voting against. The US position — that existing international humanitarian law is sufficient to govern autonomous weapons — is technically defensible and strategically disastrous. It cedes the normative high ground to nations that will write the rules in America's absence, and it provides no framework for constraining the proliferation of autonomous weapons to non-state actors, authoritarian regimes, and other actors whose compliance with international humanitarian law is neither expected nor enforceable.

The United States should lead the negotiation, not resist it. The framework should center on a principle of meaningful human control — not a ban on autonomous systems, which is neither achievable nor strategically sound, but a binding international standard that requires a human decision-maker to authorize lethal force in all but the most narrowly defined defensive scenarios. The model is the Chemical Weapons Convention, a norm that constrains the worst outcomes while preserving legitimate military options. The United States designed the CWC's verification regime and used it as a tool of American strategic advantage for three decades. It can do the same with autonomous weapons — but only if it is at the table, writing the rules rather than reacting to them.

Prescription Five: Accept the End of the Exquisite

The fifth prescription is the hardest to hear, particularly in the offices of the defense prime contractors that have shaped American military procurement for half a century. The era of the exquisite platform is ending. The F-35, at $80 million flyaway cost per airframe. The $13 billion aircraft carrier. The $2 billion destroyer. These systems are not obsolete — but they are increasingly vulnerable, and the cost of defending them against cheap autonomous threats is growing faster than the cost of the threats themselves.

The F-35 will continue to fly combat missions. The carrier strike group will continue to project power. But the next generation of military capability will not be defined by single platforms that cost billions and take decades to develop. It will be defined by thousands of attritable systems that cost thousands and take months to field. The Air Force's Collaborative Combat Aircraft (CCA) program — which aims to field autonomous wingmen that fly alongside manned fighters at a fraction of their cost — is a step in the right direction. The Replicator initiative, announced by Deputy Secretary of Defense Kathleen Hicks in August 2023 to field autonomous systems "at a speed and scale that is relevant," was another step — though its execution fell short of its ambition.

The prescription is not to cancel the F-35 or decommission the carrier fleet. It is to stop pretending that the next F-35 will be the answer to the next war. The next war will be fought with systems that cost less than the F-35's landing gear actuator. The defense establishment must plan accordingly — not in ten years, not in the next budget cycle, but now.

The Reader's Obligation

These five prescriptions are not addressed only to generals and senators. They are addressed to anyone who votes, pays taxes, or has an opinion about how the world's most powerful military spends $892.6 billion per year. The $200 war is not an abstraction confined to Ukrainian battlefields and Pentagon briefing rooms. It is a structural transformation in the economics of violence that will affect the security of every nation, the foreign policy of every alliance, and the safety of every city within range of a $200 drone — which, increasingly, is all of them.

The math is not complicated. A $200 drone can destroy a $24 million missile system. More than 40 countries have armed drones. The Pentagon spends 20 dollars on the past for every dollar it spends on the future. Venture capital has invested $49.1 billion on the bet that the future arrives before the Pentagon adjusts. The AI cliff means the cost of destruction is falling toward zero.

The $200 war is not coming. It is here. The only question is whether anyone will do the math before the math does itself.

The prescriptions are clear. The economics are undeniable. What remains is the question that economics cannot answer alone — what kind of world these numbers are building.

Conclusion: The Math

The footage is the same. A thermal camera, mounted on a small unmanned aircraft, descends toward a Russian Tor-M2 surface-to-air missile system parked at the edge of a tree line near Kupiansk, Ukraine. The Tor is worth $24 million. It was designed to shoot down cruise missiles and low-flying aircraft — threats that cost millions of dollars and arrive at the speed of sound. The drone approaching from above costs approximately $200. It was assembled in a workshop from commercially available components. It carries a shaped-charge warhead repurposed from a conventional munition. The operator, sitting in a basement several kilometers away, watches through a pair of $40 FPV goggles as the drone closes the final hundred meters. The Tor's radar — engineered to detect and track objects moving at Mach 2 — does not register the incoming threat. The drone strikes the missile system's radar array. The screen flares white. Twenty-four million dollars of air defense capability ceases to exist.

The reader saw this footage in the Introduction. The numbers were the same then. Two hundred dollars. Twenty-four million dollars. A ratio of 120,000 to 1.

But the reader who has traveled through the 15 chapters between that first viewing and this one sees different things in the footage now.

Six Centuries in Forty-Seven Seconds

The reader sees Henry V's archers at Agincourt — peasant soldiers with longbows that cost a few shillings, piercing the armor of mounted knights who represented a generation's investment in military capital. The same math. The cheap weapon, mass-produced and expendable, destroying the expensive platform that an entire economic system was organized to protect. The French knights at Agincourt did not lose because they lacked courage or skill. They lost because the economics had turned against them, and their institutions refused to adapt until the arrows were already in flight.

The reader sees the torpedo in the Philippine Sea — a $10,000 weapon, launched from a submarine that cost a fraction of its target, sinking battleships that nations had spent decades and fortunes to build. The same math. The asymmetric weapon that inverted the cost curve of naval warfare and rendered the most powerful surface combatants in the world into targets rather than projectors of force.

The reader sees the IED in Helmand Province — a $20 device buried in a road, triggering a $45 billion counter-program that employed tens of thousands of people, consumed a significant fraction of the defense budget, and never solved the problem it was created to address. The same math. The cheap weapon that the expensive institution could not defeat, not because the institution lacked resources, but because the economics of the fight guaranteed that the defender would always spend more than the attacker.

The reader sees the Houthi drone approaching a container ship in the Red Sea, and the $2 million interceptor missile launched to destroy a $2,000 threat, and the destroyer's magazine depleting itself at a rate that no logistics chain could sustain. The same math.

The cost-exchange ratio that transforms a non-state actor with a modest arsenal into a strategic threat to the world's most powerful navy.

And the reader sees what comes after the footage ends: the factory floor in Shenzhen where the drone's components were manufactured alongside smartphone parts and children's toys. The Nasdaq trading floor where Swarmer's stock price climbed from $5 to $65 on the strength of a combat-proven autonomous coordination system. The Turkish airfield where a K2 swarm drone sat in its crate, waiting for export to a country that had never possessed precision strike capability before. The Ukrainian workshop where an AI-guided drone completed its strike through a Russian jammer's static, its operator's screen black for the final eight seconds of the engagement.

The $200 drone that struck the Tor-M2 near Kupiansk was not an anomaly. It was an inflection point. Everything after it follows the math.

The Swarmer Trajectory

In Chapter 4, I traced the arc of Swarmer — the Ukrainian drone-swarm company that grew from a Kyiv basement in the early months of the Russian invasion to a Nasdaq-listed defense technology firm with a valuation that startled Wall Street analysts who had never priced a company born in a war zone. The IPO opened at $5 per share. Within two trading sessions, it surged above $65.

The financial trajectory was remarkable, but it was not the point. The point was what Swarmer represented: a company that had developed autonomous multi-drone coordination software —

the Styx system — under combat conditions, tested it in more than 100,000 missions, refined it through continuous operational feedback from front-line units, and brought it to the public markets as a proven capability rather than a speculative prototype. Swarmer was not selling a vision of autonomous warfare. It was selling a product that had been used in autonomous warfare, every day, for years.

Where Swarmer goes, the economics of warfare follow. The company's trajectory — from garage to battlefield to stock exchange — is the trajectory of the $200 war itself. The weapons start cheap. They get smarter. They prove themselves in combat. They attract capital. The capital funds the next generation of weapons, which are cheaper and smarter still. The cycle accelerates. The cost curve drops. The cost-exchange ratio widens. And the institutions that were built to fight the last war find themselves spending more and more to defend against weapons that cost less and less.

Swarmer is not unique. It is a template. The next Swarmer is already being built — in a workshop in Dnipro, in a lab in Shenzhen, in a startup accelerator in Tel Aviv, in a DARPA-funded research program in a university basement. The economics guarantee it.

But the Swarmer trajectory also carries a warning. The company's success was made possible by a war that killed hundreds of thousands of people. Its stock price climbed because its product worked, and its product worked because it was tested against human beings in conditions of extreme violence. The financial market that rewarded Swarmer with a 1,200% share price increase was not pricing the suffering embedded in the data. It was pricing the capability that the suffering validated. There is something clarifying — and something unsettling — about a stock market that assigns dollar values to combat-proven lethality. It is the $200 war in its purest form: the reduction of warfare to an economic function, optimized by algorithms, valued by traders, and measured in returns on investment.

The Census

There is one number the reader has not yet seen. It is the number that transforms the $200 war from an argument into a census.

By 2026, the total number of combat drone sorties logged worldwide exceeded 10 million.

Ten million sorties. More than the total number of manned combat sorties flown in every conflict since the Vietnam War combined — Korea, the Gulf War, the Balkans, Iraq, Afghanistan, Libya, Syria, and every smaller engagement in between. Manned combat aviation, across five decades and dozens of conflicts, generated fewer sorties than unmanned systems generated in the four years since Russia's full-scale invasion of Ukraine.

The number is not an estimate. It is a compilation: Ukrainian Ministry of Defense sortie data, which alone accounts for millions of drone flights per year on the eastern front. ACLED drone strike databases covering Africa, the Middle East, and South Asia. Defense media reporting on Russian, Turkish, Iranian, and Israeli drone operations across multiple theaters. The figure of 10 million is conservative. The actual number is almost certainly higher, because many combat drone flights — particularly those by non-state actors and by nations that do not publicly report operational data — are not captured in any database.

Ten million combat drone sorties. The $200 war is not an argument about the future. It is a description of the present.

The number rewrites the denominator in every calculation about the relevance of manned airpower. When a single conflict — Ukraine — generates more combat drone sorties in a month than the United States Air Force flew in the entire first Gulf War, the question is no longer whether unmanned systems are supplementing manned aircraft. The unmanned systems have become the primary instrument of aerial warfare, and the manned aircraft have become the supplement. The inversion happened not through a policy decision or a doctrinal shift. It happened through arithmetic. Drones are cheaper to build, faster to replace, and cheaper to lose. The math made the decision that the institutions could not.

What the Numbers Build

The five numbers from the previous chapter — $200, 120,000:1, $892.6 billion, $49.1 billion, 40+ — are not a policy brief. They are a lens. Hold them against any defense budget, any arms sale, any military procurement decision, any policy proposal, any election debate about national security, and they will tell the reader whether the people making those decisions have done the math.

A budget that spends 20 dollars on manned fighters for every dollar on autonomous systems has not done the math. A procurement program that takes seven years to field a drone the battlefield needed yesterday has not done the math. An arms control framework that regulates missiles but not the $200 weapon that has killed more armor in Ukraine than any missile has not done the math. A national security strategy that treats autonomous weapons as a future challenge rather than a present reality has not done the math.

The numbers do not prescribe a single answer. Reasonable people will disagree about the pace of budget restructuring, the design of international governance frameworks, the acceptable level of autonomy in lethal systems, and the proper balance between legacy capability and attritable mass. Those are political decisions, and they should be debated vigorously.

But the numbers are not debatable. The cost curve points in one direction. The proliferation curve points in one direction. The autonomy curve points in one direction. The math is the math.

The five numbers are also a test. When a senator proposes a $15 billion increase in fighter jet procurement, hold the numbers against the proposal and ask whether it addresses the $200 threat or the $10 million one. When a defense contractor markets a next-generation platform with a 20-year development timeline, hold the numbers against the timeline and ask whether the threat will wait. When an arms control negotiator argues that existing humanitarian law is sufficient to govern autonomous weapons, hold the numbers against the claim and ask what law governs a weapon assembled from Amazon parts by a non-state actor in a country that signed no treaty.

The Question the Money Cannot Answer

This book has followed the money — from the $200 drone to the $24 million missile system, from the $892.6 billion defense budget to the $49.1 billion in venture capital, from the Shenzhen component factory to the Nasdaq trading floor. The money tells a clear story: cheap autonomous weapons are reshaping the economics of warfare more profoundly than any development since the atomic bomb. The cost curve is permanent. The proliferation is irreversible. The autonomy frontier is advancing on a trajectory that economic incentives will not allow any nation to resist.

But the money does not answer the question that matters most. The drone revolution is not, at its core, a military story. It is an economic story. The same forces that made smartphones ubiquitous — miniaturized components, globalized supply chains, open-source software, consumer-grade manufacturing — are making precision strike ubiquitous. The same logic that put a supercomputer in every pocket is putting a guided munition within reach of every government, every militia, every group with a grievance and a wire transfer.

The question is not whether this transformation will continue. The economics guarantee it. The question is whether human judgment will survive the math. Whether the institutions built to govern the use of force — military chains of command, international humanitarian law, democratic accountability for the decision to kill — can adapt to a world where the cost of killing approaches zero and the barriers to entry disappear.

A $200 drone. A $24 million missile system. Do the math. Then do something about it.

The numbers in this book are the beginning of that work. The next book in this series, *Silicon Valley Goes to War*, tells the human story behind the numbers — the founders building autonomous weapons in converted warehouses, the military officers deciding whether to trust algorithms with lethal authority, the investors funding the companies that are arming the future, and the ethical reckoning that is coming for all of them. The money has spoken. Now it is time to meet the people spending it.

Selected Bibliography

Introduction and Chapter 1

Malyasov, Dylan. "Ukraine's Tiny Drone Blows Up Russian Tor Air Defense System." *Defense Blog*, November 9, 2023.

Hambling, David. "The Drones of Ukraine." *Forbes,* March 2024.

Oryx. Open-source intelligence project, confirmed Russian armored vehicle losses to FPV drones. Updated through February 2026.

Chapters 2 and 3

The War Zone. "FPV Drones Have Become the Leading Killer of Tanks in Ukraine." January 2026.

Sutton, H.I. "Ukraine's Maritime Drone Fleet." *Naval News,* February 2025.

South China Morning Post. "China Tests 16-Tonne Drone Mothership Capable of Carrying 100 Smaller Drones." December 2025.

Chapters 4 and 5

Government Accountability Office. "F-35 Joint Strike Fighter: Cost and Schedule Remain Challenging." GAO-24-106315, March 2024.

Congressional Research Service. "Navy Ford (CVN-78) Class Aircraft Carrier Program." CRS Report RS20643, updated January 2026.

Spinney, Franklin C. *Defense Facts of Life: The Plans/Reality Mismatch.* Westview Press, 1985.

Chapters 6 and 7

Gray, Edwyn. *The Devil's Device: Robert Whitehead and the History of the Torpedo.* Naval Institute Press, 1991.

Tooze, Adam. *The Wages of Destruction: The Making and Breaking of the Nazi Economy.* Viking, 2006.

Krepinevich, Andrew. "Get Ready for the Democratization of Destruction." *Wall Street Journal*, September 2007.

Bilmes, Linda. "The Long-Term Costs of United States Care for Veterans of the Afghanistan and Iraq Wars." Watson Institute, Brown University, August 2021.

Chapters 8 and 9

Hicks, Kathleen. Keynote address, National Defense Industrial Association Emerging Technologies Conference, August 28, 2023.

Department of Defense. "Fiscal Year 2026 Budget Request Overview." Office of the Under Secretary of Defense (Comptroller), February 2026.

Center for Strategic and International Studies. "Restructuring the Defense Budget for the Drone Age." CSIS Briefs, January 2026.

Wilson, James Q. *Bureaucracy: What Government Agencies Do and Why They Do It.* Basic Books, 1989.

Chapters 10 and 11

RAND Corporation. "Counter-Unmanned Aircraft Systems: Cost-Exchange Analysis and Investment Priorities." RAND Report RR-A2847-1, 2025.

UK Ministry of Defence. "DragonFire Laser Achieves First Successful Engagement Against Aerial Target." Press release, January 2025.

Kania, Elsa. "Battlefield Singularity: Artificial Intelligence, Military Revolution, and China's Future Military Power." Center for a New American Security, November 2017.

Financial Times. "Inside Shenzhen's Drone Component Supply Chain." March 2025.

Chapter 12

Kasapoglu, Can. "Turkey's Drone Diplomacy: Lessons for Europe." European Council on Foreign Relations, December 2022.

Africa Center for Strategic Studies. "Armed Drone Strikes Surge Across Africa." 2025.

Shaikh, Shaan, and Wes Rumbaugh. "The Air and Missile War in Nagorno-Karabakh: Lessons for the Future of Strike and Defense." Center for Strategic and International Studies, December 2020.

Chapters 13 and 14

Hambling, David. "Inside Ukraine's Drone Factories." *Forbes*, September 2024.

Scharre, Paul. *Four Battlegrounds: Power in the Age of Artificial Intelligence.* W.W. Norton, 2023.

Tucker, Patrick. "Ukrainian Drones Are Flying Themselves Into Targets." *Defense One*, March 14, 2024.

Abraham, Yuval. "'Lavender': The AI Machine Directing Israel's Bombing Spree in Gaza." *+972 Magazine*, April 3, 2024.

Chapter 15 and Conclusion

Shah, Raj M., and Christopher Kirchhoff. *Unit X: How the Pentagon and Silicon Valley Are Transforming the Future of War.* Scribner, 2024.

Gunzinger, Mark, and Lukas Auber. "Affordable Mass: The Need for a Cost-Effective PCA." Center for Strategic and Budgetary Assessments, 2024.

United Nations General Assembly. Resolution on Lethal Autonomous Weapons Systems, December 2024.

General References

Royal United Services Institute. Drone warfare analysis series, 2022-2026.

Congressional Research Service. Multiple reports on defense programs, budgets, and emerging technologies, 2024-2026.

Government Accountability Office. Defense acquisition and program assessment reports, 2024-2026.

Stockholm International Peace Research Institute. Arms Transfers Database and annual trend reports.

International Institute for Strategic Studies. *The Military Balance 2025* and *2026.* London: IISS.

Brose, Christian. *The Kill Chain: Defending America in the Future of High-Tech Warfare*. Hachette Books, 2020.

Scharre, Paul. *Army of None: Autonomous Weapons and the Future of War*. W.W. Norton, 2018.

Singer, P.W. *Wired for War: The Robotics Revolution and Conflict in the 21st Century*. Penguin, 2009.

Also by C. Scott Lannon

Hunter, Killer Series

Hunter, Killer — The definitive account of how AI drone swarms are rewriting the rules of warfare
The $200 War — How the cheapest weapon in history is rewriting the rules of power

Coming Soon

Silicon Valley Goes to War — The founders, the money, the culture clash, and the ethical reckoning of defense tech

Stay Briefed

Get the latest on AI warfare, drone technology, and defense innovation — plus early access to new books in the Hunter, Killer series.

Join the briefing list at https://cscottlannon.com/hunter-killer

Subscribers receive "The Five Drones That Changed Warfare" — a free PDF mapping five weapon systems onto the Hunter-Killer spectrum.

Next in the Hunter, Killer Series

Silicon Valley Goes to War

Hunter, Killer — Book Three

Palmer Luckey sold his VR company to Facebook for $3 billion, then walked into the California desert and started building weapons. He was twenty-four. Within a decade, his defense startup Anduril was valued at $60 billion, and a generation of Silicon Valley founders had followed him through the door — trading hoodie culture for security clearances, venture capital for Pentagon contracts, and the comfortable fiction that technology is neutral for the uncomfortable reality that someone has to build the weapons.

Silicon Valley Goes to War is the story of what happens when the fastest-moving industry on Earth collides with the slowest-moving bureaucracy. It is the story of founders who believed they could disrupt the defense industrial base the way they disrupted taxi cabs and hotel rooms — and discovered that the Pentagon does not move fast and break things. It is the story of engineers who joined defense-tech startups to build autonomous systems and then sat in ethics reviews debating whether their code should be allowed to kill without a human in the loop. It is the story of the money — billions of dollars in venture capital pouring into a sector that most VCs would not touch five years ago — and the question of what happens when the investors who funded social media are now funding the weapons that social media films.

The founders. The money. The culture clash. The ethical reckoning. And the question no one in Silicon Valley wants to answer: if you build the most powerful autonomous weapons on Earth, who decides how they are used?

Silicon Valley Goes to War is coming soon.

Made in the USA
Monee, IL
04 November 2026

67382305R00095